CELL MEMBRANES

Methods and Reviews

Volume 1

CELL MEMBRANES

Methods and Reviews

Volume 1

Edited by
Elliot Elson
William Frazier
and
Luis Glaser

Washington University School of Medicine
St. Louis, Missouri

PLENUM PRESS • NEW YORK AND LONDON

ISBN-13: 978-1-4684-4513-8   e-ISBN-13: 978-1-4684-4511-4
DOI: 10.1007/978-1-4684-4511-4

© 1983 Plenum Press, New York

A Division of Plenum Publishing Corporation
233 Spring Street, New York, N.Y. 10013

CONTRIBUTORS

A. Avivi   Department of Chemical Immunology, The Weizmann Institute of Science, Rehovot 76100, Israel

Lewis C. Cantley   Department of Biochemistry and Molecular Biology, Harvard University, Cambridge, Massachusetts 02138

William A. Catterall   Department of Pharmacology, University of Washington, Seattle, Washington 98195

I. Lax   Department of Chemical Immunology, The Weizmann Institute of Science, Rehovot 76100, Israel

T. A. Libermann   Department of Chemical Immunology, The Weizmann Institute of Science, Rehovot 76100, Israel

Ian G. Macara   Department of Biochemistry and Molecular Biology, Harvard University, Cambridge, Massachusetts 02138

Walter Neupert   Institute of Biochemistry, University of Göttingen, D-3400 Göttingen, Federal Republic of Germany

J. Schlessinger   Department of Chemical Immunology, The Weizmann Institute of Science, Rehovot 76100, Israel

A. B. Schreiber   Department of Chemical Immunology, The Weizmann Institute of Science, Rehovot 76100, Israel

Martin Teintze   Institute of Biochemistry, University of Göttingen, D-3400 Göttingen, Federal Republic of Germany

Dorothy M. Wilson   Department of Physiology, Harvard Medical School, Boston, Massachusetts 02115

T. Hastings Wilson   Department of Physiology, Harvard Medical School, Boston, Massachusetts 02115

Y. Yarden   Department of Chemical Immunology, The Weizmann Institute of Science, Rehovot 76100, Israel

PREFACE

Cell Membranes: Methods and Reviews is the continuation of a distinguished series of books edited by Edward Korn under the title *Methods in Membrane Biology*. While the original series nominally dealt with methods for the study of membranes, its chapters were, in fact, much broader and covered the conceptual framework upon which the methods were based as well as the methodology *per se*. In continuing this series, we changed the title to reflect this broader point of view. We hope to present a series of volumes published at roughly annual intervals containing comprehensive reviews dealing with various aspects of membrane biochemistry. We will on occasion also include short chapters that deal exclusively with new methods and their applications. The biology of membranes as we view it is very broad and we hope therefore not only to include work that is directly related to molecules present in membranes but also to treat molecules that interact with membrane components. In each of the volumes we hope to have one or two areas of concentration and to include chapters that collectively present a balanced and comprehensive review of those particular fields. In this first volume, we have concentrated on transport of solutes across the membrane. We are grateful to the authors, who have provided us with excellent and comprehensive chapters in these areas in a timely fashion. In subsequent volumes we shall concentrate on membrane cytoskeleton interactions and on interaction of hormones with membrane receptors and the transduction of that event into physiological effects.

To ensure rapid publication of the chapters submitted to us we will avoid postponing publication of a volume to wait for chapters that are unduly delayed. While this may leave some gaps in the presentation, the timeliness of the reviews will more than compensate. In the first volume the chapter on epidermal growth factor was to be complemented by other chapters in similar fields. Since these were not available by the deadline, we chose to publish that very excellent chapter, which can well stand on its own merits as part of this first volume, and not to delay publication to include related chapters.

We hope that this series will continue in the tradition of excellence set by Dr. Korn for the preceding volumes. The enthusiastic response and co-operation from the authors whom we have invited to write chapters makes us confident of the success of this series. Volumes 2 and 3 are now well under way and look as exciting as the present volume.

Elliot L. Elson
William A. Frazier
Luis Glaser

St. Louis, Missouri

CONTENTS

Chapter 1

Sugar–Cation Cotransport Systems in Bacteria

T. Hastings Wilson and Dorothy M. Wilson

Chapter 2

The Structure and Function of Band 3

Ian G. Macara and Lewis C. Cantley

Chapter 3

Biosynthesis and Assembly of Mitochondrial Proteins

Martin Teintze and Walter Neupert

CELL MEMBRANES
Methods and Reviews

SUGAR–CATION COTRANSPORT SYSTEMS IN BACTERIA

T. Hastings Wilson and Dorothy M. Wilson

1. INTRODUCTION

Substrate–cation cotransport systems are extremely widespread in nature, being found in bacteria (West, 1970), yeast (Eddy and Nowacki, 1971), *Neurospora* (Slayman and Slayman, 1974), *Chlorella* (Komor, 1973; Höfer *et al.,*1974), and animal cells (see reviews by Schultz and Curran, 1970; Crane, 1977). The discovery of this type of transport system resulted from early studies in animal cells which indicated the importance of sodium ion in the transport of amino acids by ascites tumor cells (Christensen and Riggs, 1952) and in the transport of sugars by the mammalian small intestine (Riklis and Quastel, 1958; Csaky and Thale, 1960; Bihler and Crane, 1962). The first explicit suggestion that sugars were cotransported with cations came from Crane (1962) and from Mitchell (1963). Crane proposed that Na^+ and glucose crossed the membrane together on the same membrane carrier during absorption by the mammalian small intestine. In 1963 Mitchell postulated that in microorganisms substrate–cation cotransport involved the proton rather than Na^+. Mitchell specifically proposed that lactose was transported by its carrier in *Escherichia coli* by cotransport with protons, the driving force for lactose accumulation being the electrochemical potential difference of protons across the membrane. Since that time a number of examples of sugar–cation cotransport in bacteria have been described. The purpose of this chapter is to review our knowledge in this field with particular emphasis on systems found

T. Hastings Wilson and Dorothy M. Wilson • Department of Physiology, Harvard Medical School, Boston, Massachusetts 02115.

in *E. coli*. The reader is referred to several previous reviews in this area (Hamilton, 1975, 1977; Simoni and Postma, 1975; Crane, 1965; 1977; Rosen and Kashket, 1978; Silhavy *et al.*, 1978; Eddy, 1978; Harold, 1978; Wilson, 1978; West, 1980).

2. LACTOSE TRANSPORT IN *ESCHERICHIA COLI*

2.1. Early Studies

The lactose transport system of *Escherichia coli* was discovered in 1956 by the group working in Monod's laboratory at the Pasteur Institute (Rickenberg *et al.*, 1956). They found that cells induced by growth on lactose (or thiogalactosides) were capable of accumulating the nonmetabolizable lactose analog thiomethylgalactoside (TMG) to concentrations far higher than that in the external medium. The intracellular sugar was in a soluble form in the cytoplasm, as shown by its osmotic activity (Sistrom, 1958) and by the fact that the sugar could be displaced quantitatively from the cell by the addition of either metabolic inhibitors or nonradioactive galactosides. Exit from the cell of galactosides occurred through the carrier since induced cells showed much more rapid exit than those that were uninduced (Kepes, 1960).

2.2. Methods for Study of Transport

Monod and his group at the Pasteur Institute utilized two general types of techniques for the study of the galactoside transport system (Rickenberg *et al.*, 1956). In the first method cells are incubated in the presence of radioactive TMG and at various time intervals the cells are separated from the medium by filtration. Following a brief wash the radioactivity on the filter is determined. This general technique is widely utilized as a transport assay for substrates that are accumulated by the cell.

A second technique involves the measurement of the *in vivo* hydrolysis of the chromogenic substrate *o*-nitrophenylgalactoside (ONPG) by intact cells (Cohen and Monod, 1975; Koch, 1969). This sugar enters the cell through the membrane carrier and is rapidly split by the intracellular β-galactosidase to free galactose, which may be metabolized, and *o*-nitrophenol, which diffuses out of the cell into the external medium, where it is readily measured because of its yellow color. The transport step is the rate-limiting process in yellow color production since β-galactosidase is present in a large excess. The ease with which this colorimetric measurement can be accomplished makes it very useful for quick assays during many types of experiments. It should be noted that metabolic inhibitors such as 2,4-dinitrophenol and azide, which completely block the accumulation of TMG, have relatively small effects on the entry rate of ONPG.

In a third technique (Maloney and Wilson, 1978) the entry rate of the natural substrate lactose may be measured in the glucose-negative mutant, GN2 (Fraenkel *et al.*, 1964). Lactose enters this strain via the carrier and is split by β-galactosidase; the glucose is quantitatively excreted, since it cannot be metabolized in this particular strain. Thus, glucose excretion (measured with a glucose oxidase reagent) may be taken as a measure of lactose entry into this cell. The primary disadvantage of this technique is the necessity for a rather high concentration of lactose (5–10 mM).

2.3. Substrates for Transport

The natural substrate for this transport system is the disaccharide lactose (glucose-4-β-D-galactoside). There is, however, a relatively low sugar specificity for the transport carrier. It is able to recognize a variety of α- as well as β-galactosides (Figure 1). The naturally occurring sugars which are substrates (in addition to lactose) include melibiose (glucose-α-galactoside), arabinose-β-galactoside, fructose-β-galactoside, and free galactose (Cohen and Monod, 1957; Sandermann, 1977; Schuldiner and Kaback, 1977). Thiogalactosides can be transported but are not hydrolyzed by β-galactosidase. These nonmetabolizable sugars can be accumulated within the cell against large concentration gradients by induced cells. A comparison of the structures of a variety of substrates is given in Figure 1.

Two substrates, thiodigalactoside (TDG) and *p*-nitrophenyl-α-galactoside (α-PNPG), have very high affinity for the carrier and thus are useful as inhibitors of transport or for protection of the sugar binding site. Thiodigalactoside has been used by Kennedy and his collaborators to protect the reactive sulfhydryl group of the lactose binding site against inhibition by *N*-ethylmaleimide (NEM) (Fox and Kennedy, 1965; Jones and Kennedy, 1969).

GALACTOSIDE	AGLYCONE	NAME

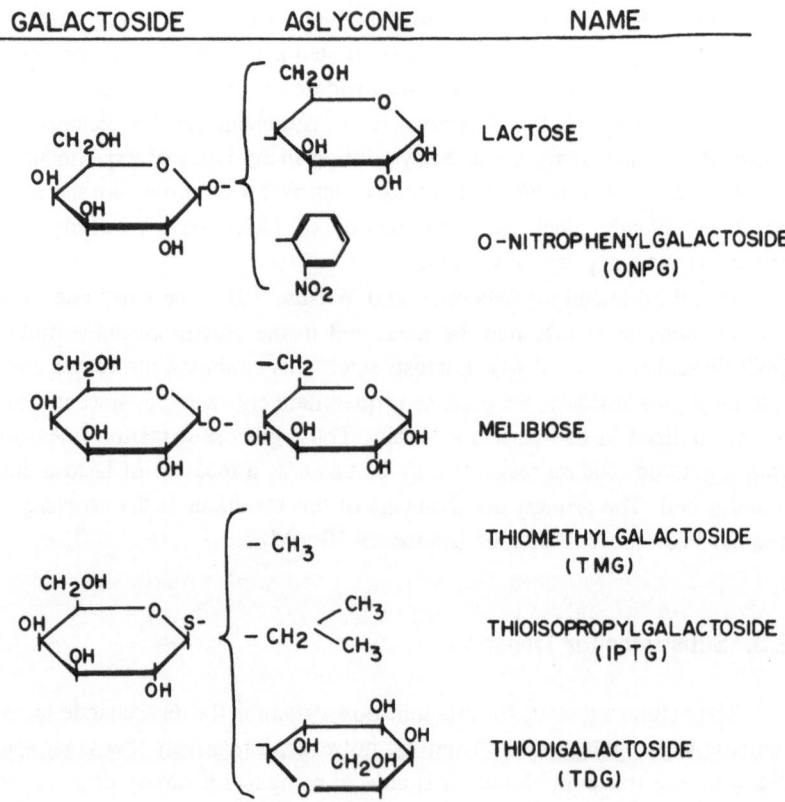

FIGURE 1. Substrates for the lactose transport system of *Escherichia coli*. (From Wilson, 1978.)

2.4. Lactose–H⁺ Cotransport

According to the cotransport hypothesis (Mitchell, 1963) there is obligatory coupling between proton and sugar translocation. The first direct demonstration of proton movement associated with lactose transport came from West (1970). He showed that the addition of lactose to energy-depleted cells resulted in loss of protons from the external medium and acidification of the interior of the cell (West, 1973). In the complementary experiment it has been shown that imposition of an inwardly directed protonmotive force resulted in galactoside accumulation. Thus TMG accumulation was observed in energy-depleted cells by producing a membrane potential (Hirata *et al.*, 1974) or a pH gradient (Flagg and Wilson, 1976).

2.5. Chemical Identification of the Lactose Carrier

The first chemical identification of the membrane protein component of the lactose transport system was made by Fox and Kennedy in 1965. They devised an ingenious technique for labeling the membrane carrier. This method is based upon the fact that certain galactosides protect the essential SH group of the lactose carrier from the binding by NEM. In this technique cells are first incubated in the presence of TDG to protect the SH group of the transport site. This is followed by the addition of nonradioactive NEM. During this period the NEM reacts with many proteins other than the lactose carrier. The cold NEM and TDG are then washed from the cell membranes and [^{14}C]-

TABLE I. Genetic Control of the M Protein[a]

Strain	Transport activity	Protein-bound NEM (pmoles/mg protein)		M protein (pmoles/mg protein)
		No TDG	10 mM TDG	
ML-308 ($I^-Z^+Y^+$) (constitutive)	+	197	86	111
ML-30 ($I^+Z^+Y^+$) (induced)	+	237	107	130
ML-30 ($I^+Z^+Y^+$) (uninduced)	−	124	125	—[b]
ML-35 ($I^-Z^+Y^-$) (constitutive)	−	93	96	—[b]
ML-308-225 ($I^-Z^-Y^+$) (constitutive)	+	252	129	123
ML-3 ($I^+Z^+Y^-$) (induced)	−	99	100	—[b]

[a] From Fox *et al.* (1967).
[b] Not significant.

NEM is added. The radioactive compound reacts with the SH group in the lactose carrier, which is no longer protected by the sugar. With this assay Fox *et al.* (1967) studied the content of lactose carrier in a variety of strains of *E. coli*. As can be seen from Table I, the carrier could be identified in induced or constitutive strains. It was absent from uninduced cells as well as the *lacY⁻* strain ML-3.

The carrier protein was found to be firmly bound to the membrane fraction of the cell and could be extracted only with detergents such as Triton X100. After labeling the lactose carrier with radioactive NEM Jones and Kennedy (1969) extracted the membranes with sodium dodecyl sulfate, (SDS) and fractionated the extract on a column of Sephadex G-150. As shown in Figure

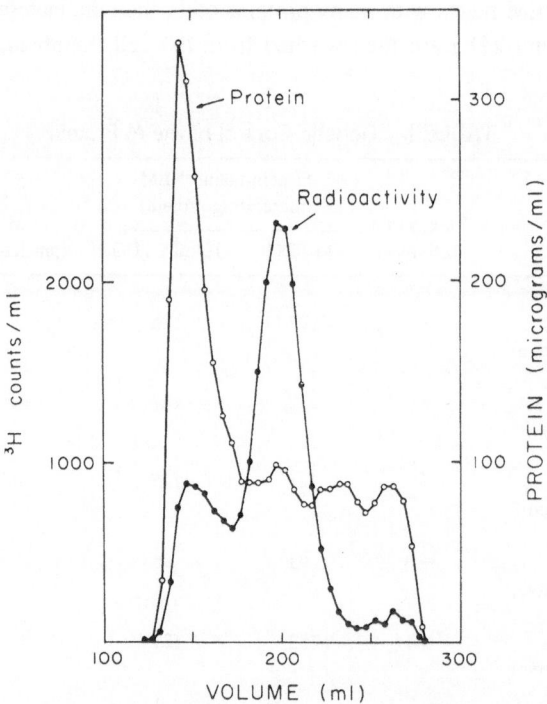

FIGURE 2. Chromatography of labeled lactose carrier on Sephadex G-150 in the presence of SDS. The carrier protein in cell-free particulate fractions was specifically labeled with tritiated NEM, extracted with buffer containing 1% SDS, and chromatographed on a column of Sephadex G-150 (100 × 2.5 cm) previously equilibrated with 1% SDS. Samples of 5 ml each were collected and assayed for total protein and radioactivity. (From Jones and Kennedy, 1969.)

2, most of the total counts were found in a single peak at the position expected for a protein of molecular weight 31,000. A considerable degree of purification was attained in this experiment, although a homogeneous protein was not obtained. No further progress in purification was made until a plasmid-containing strain that overproduced the lactose carrier was made available.

2.6. Amplification of the Lactose Carrier Protein

An important advance was made by Teather *et al.* (1978) in the isolation of a hybrid plasmid carrying the *Y* gene. Cells carrying this plasmid produce approximately ten times as much lactose transport protein as the haploid parental cells. The authors state that in the plasmid strain the lactose carrier "comprises 15% of the cytoplasmic membrane protein synthesized in the first generation after induction, compared with a wild type strain induced under the same conditions, where lactose carrier protein comprises 1.4% of the cytoplasmic membrane protein." Some of the properties of this strain are shown in Figure 3.

Strain T31RT consists of T28RT $[I^+O^+Z^+Y^-(A^+)/F'I^qO^+Z^+Y^-(A^+)]$ carrying the plasmid pC7. The plasmid is a hybrid of pBR322 and the lactose operon $(I^+Z^-Y^+A^+)$. When induced with isopropylthiogalactoside (IPTG) this cell produces β-galactosidase, transport activity, and transacetylase. Shortly after incubation with IPTG the transacetylase activity is about ten times that of T44RT while the transport activity is elevated only fourfold. TDG binding data suggest that the lactose carrier protein is probably elevated tenfold. Apparently the protein is inserted into the membrane normally but fails to show full transport function. The reason for the low carrier activity is not yet fully understood. Perhaps the energy available is insufficient for elevated transport activity. This amplified strain and others derived from it were used for purification of the lactose carrier, as will be discussed in Section 2.11.

2.7. DNA Sequence

In 1980, valuable fundamental information was reported by Büchel *et al.*, who have determined the DNA sequence of the lactose permease gene. A DNA fragment containing the LacY gene was obtained by restriction en-

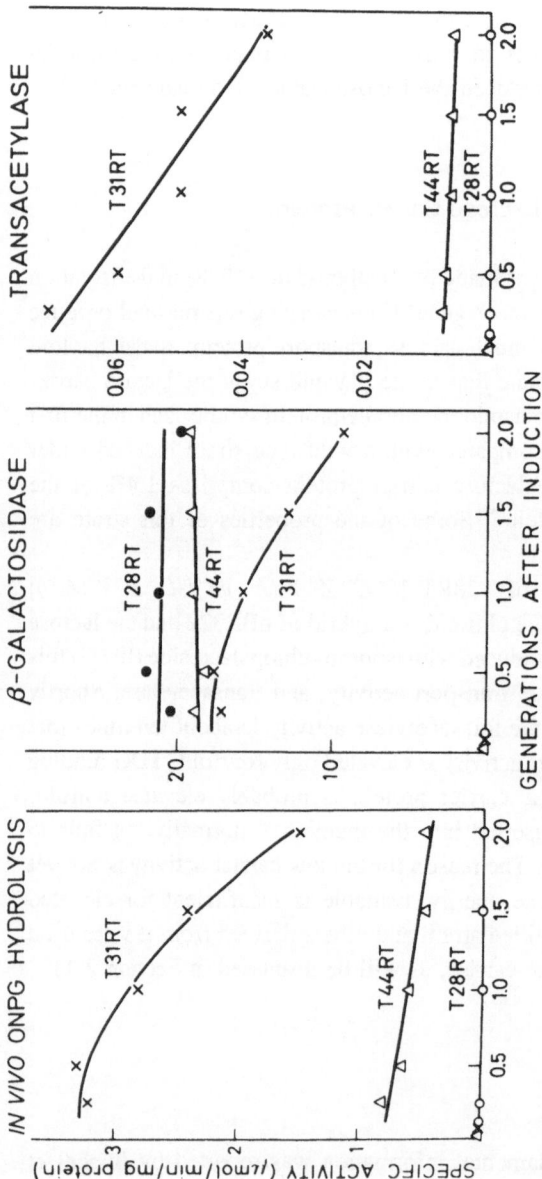

FIGURE 3. Induction of the lac operon in strains containing a lac plasmid. Exponentially growing cultures in CR minimal medium were induced with 0.5 mM IPTG at OD_{420} 0.3. Samples were taken at the indicated growth increments and the cells prepared for the assay of *in vivo* ONPG hydrolysis (lactose carrier activity), β-galactosidase, and transacetylase. The data are expressed as μmoles per minute per mg cellular protein synthesized from the time of induction to the sampling time. All assays were conducted at 28°C. The genotypes are as follows:

lac-genotype = chromosome/F'/plasmid

T31RT   = $I^+O^+Z^+Y^-(A^+)/I^qO^+Z^+Y^-(A^+)/I^+O^+\Delta(Z)Y^+A^+$

T28RT   = $I^+O^+Z^+Y^-(A^+)/I^qO^+Z^+Y^-(A^+)/-$

T44RT   = $I^+O^+Z^+Y^-(A^+)/I^+O^+\Delta(Z)Y^+A^+/-$

(From Teather *et al.*, 1978.)

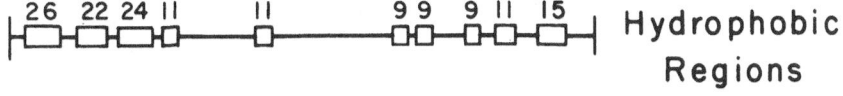

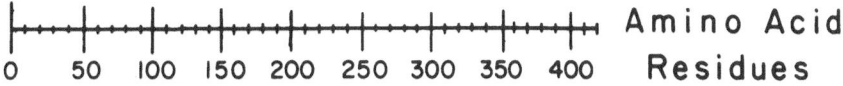

FIGURE 4. Hydrophobic and hydrophilic sequences of the lactose carrier. (From Mieschendahl *et al.*, 1981.)

donuclease digestion with Eco R_1 and cloned in the single-stranded DNA phage M13mp2. The DNA sequence of the LacY gene was determined. The amino acid sequence of the LacY gene product has a number of interesting features. It is strongly basic with a calculated isoelectric point of 10.2. The protein consists of 417 amino acids with a molecular weight of 46,504. It has a high content of nonpolar residues (71%) with a remarkably high content of phenylalanine (56 out of 417). Of the 56, 10 are found in the lipophilic region containing 26 amino acids (9–34) near the N terminus. Büchel *et al.* (1980) conclude that the molecule may be divided into four structural regions: an N-terminal region with four hydrophobic sequences, two hydrophilic regions in the middle part of the protein separated by a lipophilic sequence of 11 amino acids, and a C-terminal region with several hydrophobic sequences (Figure 4). In a subsequent publication from the same laboratory, Mieschendahl *et al.* (1981) state: "We propose that the hydrophobic regions at the N-terminus and the C-terminus act as anchors by which the protein is fixed in the membrane; they could also form pores for galactoside and proton transport."

2.8. Is There Processing of the Lactose Carrier?

Since some membrane proteins, especially those secreted across lipid bilayers, are processed by proteolytic cleavage, it was of interest to evaluate this possibility in the lactose carrier. Experiments by Villarejo (Villarejo and Ping, 1978; Villarejo, 1980) and by Fried (1981) suggested the possibility of processing of the *Y* gene product. However, recent studies of Ehring *et al.* (1980) using the overproducing plasmid pGM21 provide strong evidence against posttranslational modification. Ehring *et al.* (1980) used Triton X-100 to purify the protein from the membrane of the plasmid-containing cell. The protein subsequently aggregated and could be removed by centrifugation. The precipitate was then washed free of contaminating proteins. Ehring *et al.* (1980) determined the N-terminal 15 amino acids of the lactose carrier produced *in vivo* and that synthesized *in vitro*. The amino acid sequences in these two cases were the same and agreed with those predicted from the DNA sequence. This indicates that there is no cleavage of a "signal sequence" during the insertion of the protein. Since there is good agreement between the relative proportions of the amino acid composition of the purified carrier and that predicted from the DNA sequence, processing at the C-terminal end of the molecule seems unlikely.

The apparent molecular weight of the carrier of 31,000 determined by SDS–polyacrylamide gel electrophoresis (Jones and Kennedy, 1969; Teather *et al.*, 1978; Ehring *et al.*, 1980; Newman *et al.*, 1981) is very different from the 46,000 value obtained from the sequence data. However, Beyreuther *et al.* (1980) have reported that when the percentage of acrylamide in the SDS–polyacrylamide gel is increased from 10% to 20% the apparent molecular weight from the gel increases from 31,000 to 46,000. Thus it appears that the apparent molecular weight determined in SDS with 10% acrylamide gel is lower than the true value, probably owing to high SDS binding by this hydrophobic protein.

2.9. The SH Groups of the Carrier Protein

Beyreuther *et al.* (1981) have identified the cysteine residue at position 148 as the TDG-protectable SH group that reacts with NEM in intact cell membranes. It is presumed that this group is part of the sugar recognition site. Seven other SH groups were found in detergent-extracted protein (Beyreuther and Ehring, 1980). No disulfide groups were detected.

2.10. Solubilization and Reconstitution of the Lactose Carrier

Newman and Wilson (1980) have described a technique for the solubilization of the lactose carrier with octylglucoside and the reconstitution of transport activity utilizing the octylglucoside dilution procedure described by Racker *et al.* (1979). The *E. coli* strain used for these studies was ML-308-

225 (Winkler and Wilson, 1966), which is $LacI^-Z^-Y^+$. During extraction from the membrane the protein must be stabilized by added phospholipid. Several types were tested and a purified phospholipid fraction from *E. coli* was most effective in stabilizing the carrier. The residual membrane was removed by centrifugation. The supernatant was mixed with sonicated *E. coli* phospholipid and diluted 50-fold in potassium phosphate buffer, following the procedure of Racker *et al.* (1979). The proteoliposomes which formed following the dilution of the detergent were collected by centrifugation. These proteoliposomes were then resuspended and exposed to radioactive lactose for assay of transport.

One type of assay takes advantage of the counterflow phenomenon. Proteoliposomes are preloaded with 20 mM nonradioactive lactose, centrifuged, and then exposed to a low concentration of [^{14}C]lactose. Radioactive molecules entering proteoliposomes are temporarily trapped inside since the

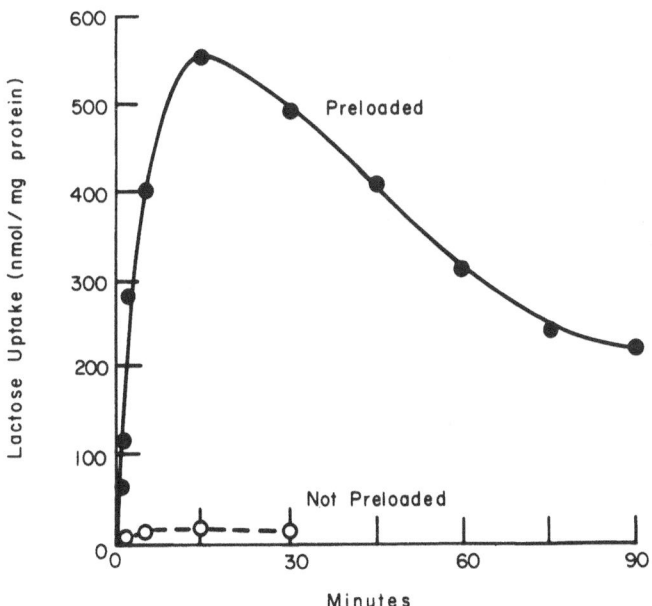

FIGURE 5. Lactose counterflow in proteoliposomes. Proteoliposomes containing lactose carrier were prepared in the presence of 20 mM lactose (preloaded) or in its absence (not preloaded), centrifuged, and resuspended in a final concentration of 0.4 mM [^{14}C]lactose. At various times samples were filtered on a 0.22-μm Millipore filter washed with buffer and the filter counted. (Redrawn from Newman and Wilson, 1980.)

nonradioactive molecules occupy most of the sites for exit. This competition for exit declines gradually as the nonradioactive molecules leave the proteoliposomes, finally permitting the accumulated [^{14}C]sugar to exit on the carrier. Figure 5 shows that accumulation proceeds for 15 min, followed by a slow decline. In the absence of preloading the uptake of [^{14}C]lactose is very much less than in the preloaded proteoliposomes.

A second type of assay is the lactose accumulation induced by a protonmotive force. Proteoliposomes containing potassium phosphate are exposed to sodium phosphate plus [^{14}C]lactose. After the lactose has equilibrated, valinomycin is added. This K$^+$ ionophore permits K$^+$ to exit from the proteoliposomes, generating a membrane potential (inside negative). The membrane potential provides the driving force for proton entry, which results in lactose accumulation (Figure 6).

Another procedure for the reconstitution of the lactose carrier has been reported by Wright *et al.* (1982). The lactose carrier was extracted with a combination of Lubrol PX and dodecyl-maltoside from cytoplasmic membranes derived from carrier-overproducing strains. In the presence of these

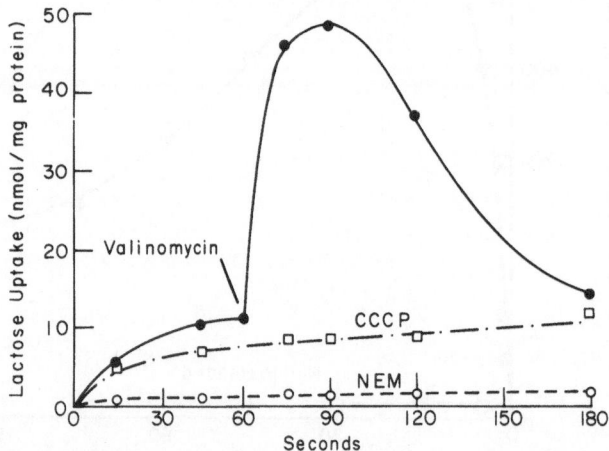

FIGURE 6. Membrane-potential-driven lactose transport in reconsituted proteoliposomes. Potassium phosphate-containing proteoliposomes with lactose carrier were diluted into sodium phosphate plus [^{14}C]lactose. Uptake of lactose was followed for 60 sec. Valinomycin (14 μM final concentration) was added and measurements continued for an additional 2 min. The proton conductor CCCP (20 μM final concentration) was added in one experiment and NEM (4 mM) was added in another. (Redrawn from Newman and Wilson, 1980.)

detergents the carrier retained its ability to bind galactosides. Addition of *E. coli* phospholipids and removal of detergents resulted in proteoliposomes with galactoside binding activity. After a freeze–thaw step countertransport of dansylgalactoside could be demonstrated in lactose-loaded proteoliposomes. Artificial membrane potentials or pH gradients did not drive sugar accumulation, possibly due to a high proton permeability of the proteoliposomes (Wright *et al.*, 1982).

2.11. Purification of the Carrier in an Active Form

Purification of the lactose carrier in an active form was reported by Newman *et al.* (1981). Membranes from the plasmid-containing strain T206 (Teather *et al.*, 1980) were first extracted with 4 M urea and subsequently with 6% sodium cholate to remove extraneous proteins (Padan *et al.*, 1979b). The residual membrane was then extracted with octylglucoside, which removed the lactose carrier from the membrane. This octylglucoside extract

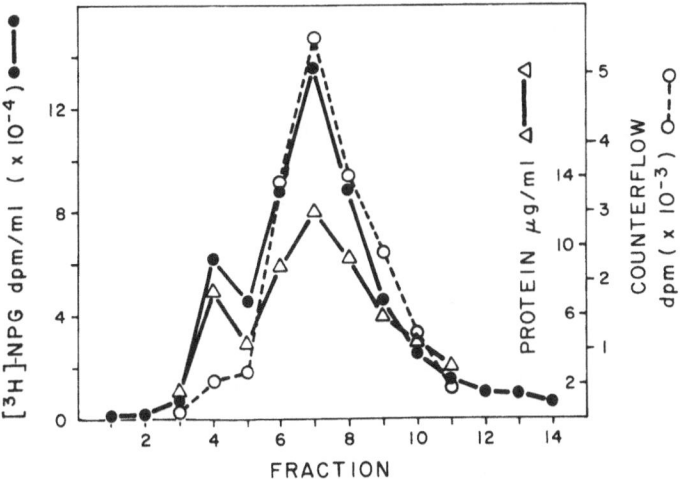

FIGURE 7. Purification of the lactose carrier by chromatography on a DEAE–Sepharose column. An octylglucoside extract of a urea–cholate-extracted membrane preparation of strain T206 (Teather *et al.*, 1980) was placed on a 10-g column of DEAE–Sepharose. The column buffer contained 1.25% octylglucoside and 1 mg/ml *E. coli* phospholipid. Protein, [3H]-photoaffinity-labeled lac carrier, and reconstituted transport activity were measured in various fractions. (From Newman *et al.*, 1981.)

was passed over a DEAE–Sepharose column, which removed virtually all of the impurities. The protein that was eluted from the column under these conditions was the lactose carrier. The first portion of the protein comes out in the void volume and appears to be an aggregated form of the carrier. This is followed by a major peak, which is pure lactose carrier protein and which possesses transport activity when reconstituted into proteoliposomes (Figure 7).

2.12. Kinetics

In the absence of protonmotive force the affinity of the carrier for lactose (or TMG) is greatly reduced (Manno and Schacter, 1970; Kaczorowski and Kaback, 1979a,b; Page and West, 1980; Wright et al., 1981). The latter three groups agree that the K_m of uptake of lactose is of the order of 15 mM in the unenergized state and about 0.2 mM when energized. The presence of only a slight protonmotive force gives K_m values close to the maximally energized level (Page and West, 1980; Robertson et al., 1980). This probably accounts for the failure of earlier workers (Winkler and Wilson, 1966; Lancaster et al., 1975) to observe the change in K_m of entry on energization.

The complete kinetic parameters for lactose entry and exit were measured by Kaczorowski and Kaback (1979a,b) for membrane vesicles of the strain ML-308-225 (Winkler and Wilson, 1966). Table II shows that energization of the membrane with a protonmotive force not only reduces the K_m for entry (from 18.9 to 0.17 mM) but also increases the V_{max} for entry and decreases the V_{max} for exit. One surprising finding was the tenfold difference in the K_m of entry compared to the K_m of exit in the unenergized state. This asymmetry would be expected to result in a lower steady-state concentration of lactose inside than outside in energy-depleted cells. This type of "reverse" sugar gradient has never been observed in energy-depleted cells or membrane vesicles; further studies of the kinetics of such cells would be useful.

A careful study comparing the carrier binding constant (K_D) of galactosides to transport affinity (K_m) was made by Wright et al. (1981). They measured binding of galactosides by the lactose carrier in membrane vesicles derived from E. coli strains of T185 or T206 which contain plasmids bearing the LacY gene. Several of these substrates, including lactose, show a very low affinity for binding to the carrier (Table III). The K_D for lactose is 15 mM, while the half-saturation constant for active transport (K_m) of cells is

TABLE II. Kinetics of Lactose Transport in Membrane Vesicles[a]

	K_m (entry) (mM)	K_m (exit) (mM)	V_{max} (entry) (nmoles/min per mg protein)	V_{max} (exit) (nmoles/min per mg protein)
Not energized	18.9	2.1	53	53
Energized	0.17	2.1	115	30

[a] Data obtained by Kaczorowski and Kaback (1979b) in membrane vesicles of ML-308-225 (*lacZ⁻Y⁺*), pH 7.

0.27 or 56 times smaller. A similar discrepancy between K_D and K_m was observed for TMG and β-ONPG galactoside. On the other hand, a second group of substrates including TDG and α-PNPG shows binding constants very similar to the K_m for transport. These findings help to explain the previous observation by Kennedy and his collaborators that lactose and certain other substrates did not protect the carrier against inactivation by *N*-ethylmaleimide, while other substrates including thiodigalactoside did protect. On the basis of these experiments they originally postulated two separate sites for substrate binding, only one of which was close to the NEM binding site (Kennedy, 1970; Kennedy *et al.*, 1974). However, in the NEM protection experiments lactose was used at a concentration (5 mM) below the K_D, and therefore the sugar would not be expected to give protection. TDG on the other hand was used at concentrations higher than the binding constant (Fox and Kennedy, 1965; Carter *et al.*, 1968; Kennedy *et al.*, 1974). It thus appears that all observations can be explained by the assumption of a single binding site for galactosides in the lactose carrier.

TABLE III. Binding Constant (K_D) Compared with Transport K_m[a]

Sugar	Binding K_D (mM)	Transport in cells K_m (mM)	Ratio K_D/K_m
Lactose	15	0.27	56
TMG	42	0.87	48
β-ONPG	15	0.91	16
α-PNPG	0.022	0.024	1
TDG	0.053	0.044	1
Melibiose	0.51	0.5	1
Raffinose	0.98	0.89	1

[a] From Wright *et al.* (1981).

If the striking difference between the K_m and K_D reported by Wright *et al.* were fundamental to the mechanism of active transport, then one might predict that lactose and TMG would be accumulated and that TDG and α-PNPG would not. However, Herzenberg (1959) has demonstrated a 40-fold accumulation of TDG by induced cells of ML-30. Accumulations of α-PNPG by *E. coli* of 30-fold have also been reported (Putzrath and Wilson, 1979). Kaczorowski and Kaback have studied TDG transport and found that the K_m for entry into the unenergized cell (facilitated diffusion) was 5 mM, which is far greater than the K_D (0.05 mM). In the energized state the K_m and K_D are similar in magnitude. Thus there is no simple relationship between K_m and K_D.

Wright *et al.* believe that galactoside and proton binding are independent processes in the absence of a protonmotive force. The binding of lactose and also of α-PNPG is not markedly influenced by the external pH in the range of 5–8. Kennedy *et al.* (1974) had previously demonstrated that TDG binding was also not strongly pH dependent. On the other hand distinct effects of pH may be observed on some aspects of the transport process. Kaczorowski and Kaback (1977a,b) found that the maximum rate of [^{14}C]lactose efflux from preloaded membrane vesicles increased about threefold with external pH values varying from pH 5.5 to 7.5. In the same study, exit showed no pH sensitivity in the presence of external unlabeled lactose and in addition was ten times faster under these conditions. Booth and Hamilton (1980) also found that exit of lactose from cells was stimulated fourfold by increasing the external pH from 5.9 to 8.1. The most straightforward explanation for the effects of pH on exit is that protonation of the unloaded carrier at the outer surface slows the return of the carrier to the inner surface and thus slows net efflux. However, when a high concentration of nonradioactive lactose is present on the outer surface, reentry of the H$^+$–sugar-loaded carrier is rapid, and it follows that exit of the labeled sugar also is rapid. West (1980) found that acidic external media slowed the efflux of TMG from whole cells deenergized with a proton-conducting ionophore. In addition he noted that this inhibition by acid media was considerably less in the "uncoupled" mutant isolated by Wong *et al.* (1970), which is unable to accumulate galactosides and which shows no net proton uptake during transport. He suggested that the binary complex of the protonated carrier had become mobile in this mutant. The commonly considered model for cation–substrate cotransport proposes that the two binary complexes (carrier–H$^+$ and carrier–substrate) do not cross the membrane or do so very much more slowly than the empty carrier or the ternary complex (carrier–H^+–substrate).

A model in accordance with this proposal that has been suggested by Wright *et al.* (1979) is the following: The unloaded carrier at the inner surface reorients across the membrane at a frequency of P_X; then at the external surface it binds to substrate and protons. The ternary complex reorients at a frequency of P_C. According to Wright *et al.* the protonation of the carrier at the external surface of the cell as a result of the protonmotive force results in an increased affinity of the carrier for the sugar. In addition they suggest a lowering of the transport K_m owing to the greater mobility of the complex compared to the unloaded carrier (P_C is greater than P_X). With regard to the latter point, Robbie and Wilson (1969) found that the P_C was three times greater than P_X, while Kaczorowski and Kaback (1979a,b) found lactose efflux to be ten times greater during exchange reactions (in membrane vesicle preparations). The mechanism by which a membrane potential (the major form of protonmotive force) can affect the affinity of the carrier may be explained by Mitchell's "proton well" hypothesis (Mitchell, 1968, 1976, 1977; Mitchell and Moyle, 1974). The "proton well" is a proton-conducting cleft or well in the membrane. Protons are attracted inward by the electric field to the bottom of the well; membrane potential is thus converted into a pH gradient. The pH at the bottom of the well then directly affects the carrier. Such a mechanism has been invoked by Schwab and Komor (1978) to account for accumulation of glucose in *Chlorella*.

2.13. Effect of $\Delta\bar{\mu}_{H^+}$ on the Structure of the Carrier

Substances that react with sulfhydryl groups are known to inhibit lactose transport strongly (Kepes, 1960; Fox and Kennedy, 1965). Recently Cohn *et al.* (1981) have shown that energizing the membrane increases the rate of inactivation by various maleimides. This is true not only of the lactose carrier but also of the melibiose and proline carriers. A similar effect of energization on phosphate transport by mitochondria has been reported by LeQuoc *et al.* (1977). Energizing with a protonmotive force probably causes an allosteric change in the carrier protein to a form in which the SH group is more sensitive to the maleimides.

Another effect of protonmotive force on conformation is suggested by

Padan et al. (1979a). These authors have found that the histidine-alkylating reagent diethylpyrocarbonate (DEPC) blocks $\Delta\bar{\mu}_{H^+}$-driven lactose transport and that this inhibition is threefold more severe in the presence of respiratory substrates. DEPC apparently has little or no effect on facilitated diffusion. The authors conclude that the histidine residue(s) alkylated under these conditions may be involved in the binding or translocation of protons.

2.14. Mutants of the Lactose Carrier

Two types of mutants have been isolated, those with complete loss of function and those with functional alterations. The most frequently encountered mutant of the latter type is a cell that grows on lactose only in the presence of the nonmetabolizable inducer IPTG (Kusch and Wilson, 1973; Langridge, 1974; Flagg and Wilson, 1976; Hobson et al., 1977; Fried, 1977, 1981; Mieschendahl et al., 1981). Apparently these mutants fail to accumulate lactose sufficiently well to induce the lactose operon, but when full induction is provided by IPTG, lactose entry is adequate for growth.

Two mutants (ML-308-22 and X71-54) have been isolated from lac constitutive strains and show a severe defect in accumulation of galactosides but are unimpaired for carrier-mediated entry (facilitated diffusion) of sugars (Wong et al., 1970; Wilson et al., 1970). The mutant ML-308-22 grows normally on lactose at concentrations of 25 mM but fails to grow at 0.25 mM, presumably owing to its inability to accumulate lactose. West and Wilson (1973) demonstrated that these mutants show a severe defect in the transport of protons with galactosides. It is concluded that the proton recognition site is abnormal, while the sugar receptor site is intact.

A more severely defective cell of a similar type was isolated by Fried (1977, 1981). This cell (3000X19), which he designated LacY^f, grows on lactose only in the presence of IPTG. It shows a complete inability to accumulate substrates against a concentration gradient but shows normal carrier-mediated β-ONPG entry. Fried (1981) showed that his LacY^f mutant accumulates a novel soluble polypeptide with a molecular weight of 87,000 which appears to be a precursor of the mutant lactose transport protein. It is proposed that this precursor is a protein chimera containing both LacY and LacA gene products.

The possibility of two distinct cistrons in the Y gene, one for sugar recognition and one for accumulation, was investigated by Langridge (1974).

He isolated and mapped 69 Y gene mutants, which were crossed with episomal deletions that divided the gene into 13 regions. He showed that the Y gene consists of only one cistron coding for a peptide capable of both facilitated diffusion and active transport.

Saier *et al.* (1978) have isolated LacY mutants that are resistant to growth inhibition by α-methylglucoside (α-MG) or glucose. Normal cells, induced for the glucose phosphotransferase (PTS) system, fail to grow on lactose when a low concentration of glucose or αMG is added. It is presumed that the α-MG-resistant mutants possess an abnormal regulatory site that frees them from control by elements of the PTS system.

Hobson *et al.* (1977) isolated 535 LacY mutants and crossed them with 48 deletions, which divided the gene into 36 deletion groups. Since there was a good correlation between the growth of mutants on lactose or on melibiose (at 42°, where the melibiose carrier is inactive) they found no support for the suggestion by Carter *et al.* (1968) of two distinct sugar binding sites. They mapped the Y^{UN} mutant (X71-54) of Wilson *et al.* (1970) and found it to lie in deletion group XXXVI or to be a deletion between the LacY and LacA genes.

Mieschendahl *et al.* (1981) have reported mutants that help to define the functional organization of the lactose carrier. They screened their collection of Y^- mutants (Hobson *et al.*, 1977), which fail to grow in 5 mM lactose, and found that 18 of them would grow if the concentration of lactose was increased to 100 mM. These apparently had a mutation in the sugar recognition site that reduced its affinity for lactose. It is interesting that all except one of these mutants map between codons 191 and 360. In addition these authors used a strain with a deletion in the maltose transport system (Shuman and Beckwith, 1979) to isolate cells that would grow on 5 mM maltose plus IPTG. In such mutants the lactose carrier transports maltose, a sugar not recognized by the normal carrier. The mutants obtained could not be mapped as they still transported lactose and melibiose; however, one Y^- mutant that had previously been mapped at codon 265 showed the ability to grow on maltose. These mutants represent another example of an altered sugar recognition site. Four mutants were isolated that show negative dominance when inserted into a strain containing a functional lac carrier of low expression. The authors infer from such mutations that the lactose carrier is active as a dimer or oligomer.

These authors also isolated cells with LacZ–Y fusions (Mieschendahl *et al.*, 1981). Deletions were obtained in which the N terminus of the lactose carrier was replaced by residues of the enzyme β-galactosidase. In spite of this the Y gene product was normally active, and thus neither the intercistronic region between the Z and Y genes nor the first few amino acids of the N-terminal region of the Y gene are necessary for transport activity.

2.15. Other Mutants Affecting Lactose Transport

Hong (1977) has isolated temperature-sensitive mutants of $E.$ $coli$ that show a defect in transport of several proton–substrate cotransport systems. The mutants map in a region (minute 65) far removed from that of the LacY gene (minute 8). He has proposed that there is an energy-coupling protein required for coupling protonmotive force to the Y gene product (as well as other transport proteins). Plate and Suit (1981) have also isolated mutants (mapping at minute 87) that show reduced transport in a variety of proton cotransport systems. The explanation for the mutants isolated by Hong and by Plate and Suit is not yet clear. Since the pure lactose carrier is capable of transport without other proteins, the role of a "proton-coupling" protein might perhaps be of a regulatory nature.

3. GALACTOSE TRANSPORT (GalP)

3.1. *E. coli*

A superabundance of pathways exists by which galactose may be transported into *E. coli* (see reviews by Kornberg, 1976; Silhavy *et al.*, 1978). This sugar can enter the cell not only by means of the lactose and melibiose carriers (Rotman *et al.*, 1968) but also by two arabinose-induced systems (Novotny and Englesberg, 1966; Daruwalla *et al.*, 1981), two galactose-induced pathways, and the glucose phosphotransferase system (Kornberg and Riordan, 1976; Postma, 1976). The clear separation of these pathways physiologically and genetically has taken many years.

The first demonstration of active transport of D-galactose was by Horecker *et al.* (1960a,b) using a galactokinaseless *(galK)* strain of *E. coli* ML. The sugar was accumulated against a concentration gradient and could be displaced by addition of 2 : 4 dinitrophenol. Galactose transport was constitutive in the *galK* mutants although the parent was inducible for the property. It was later shown by Rotman and Radojkovic (1964) that in K_{12} strains of *E. coli* two K_m values could be obtained for [^{14}C]galactose uptake by *galK* strains, and thus that two carriers were involved. In 1965 Ganesan and Rotman isolated a mutant of a *lac*-deleted K_{12} strain in which one of the two systems (MglP) was missing. D-Galactose was accumulated by this mutant by a system designated as GalP. Later work by Rotman *et al.* (1968) demonstrated that GalP could be induced by TMG, D-galactose, and D-fucose. Wilson (1974) studied the GalP transport system in strains of *E. coli* lacking MglP and LacY and showed that D-fucose was a substrate in addition to galactose. He also showed that a binding protein was not necessary and that the system was under the control of the galactose regulator gene.

The cotransport of protons with galactose was first demonstrated by Henderson (1974). The addition of galactose or D-fucose to anaerobic cell suspensions of wild type *E. coli* (induced with galactose) caused alkalinization of the suspending medium. These cells were grown at 37°, eliminating the participation of the melibiose carrier, and IPTG failed to induce H^+ movement, excluding the LacY transport system. A definitive demonstration of the involvement of protons in transport of sugars by the GalP pathway (and not by the MglP system) was presented by Henderson *et al.* in 1977. MglP⁻GalP⁺ and MglP⁺GalP⁻ strains were constructed in *galK E. coli*. When GalP was present, anaerobic proton transport occurred on the addition of galactose or fucose to the cells. Additional substrates for the GalP carrier are glucose, 2-deoxyglucose, and 2-deoxygalactose (Henderson *et al.*, 1977; Henderson and Giddens, 1977). After many years of study it is now clear that K_{12} strains of *E. coli* grown on galactose induce a proton-cotransport system (GalP) that transports galactose with a relatively high affinity ($K_m = 4 \times 10^{-6}$ M).

3.2. *Streptococcus lactis*

The galactose transport system of *Streptococcus lactis* has been the subject of a number of investigations. Desai and Goldner (1969) showed that the nonmetabolizable sugar TMG was accumulated by galactose-grown cells of *S. lactis*. Sugars that possess a significant affinity for the carrier include galactose, TMG, IPTG, TDG, methyl-α-galactoside, and methyl-β-galactoside (Desai and Goldner, 1969; Kashket and Wilson, 1972a; Thompson, 1980). One great advantage of this cell for the study of energy coupling mechanisms is that energy depletion can be accomplished simply by removal of fermentable substrate. An additional feature of use in transport studies is its sensitivity to valinomycin, the potassium ionophore that may be used to produce an artificial membrane potential. Kashket and Wilson (1972b) added valinomycin to energy-depleted cells suspended in K^+-free incubation medium containing TMG and observed sugar accumulation. The K^+ diffusion potential resulted in a membrane potential (inside negative), providing the driving force for H^+ entry. Proton entry via the galactose carrier resulted in sugar accumulation. An artificial pH gradient (outside acid) also resulted in TMG accumulation. Furthermore TMG addition to energy-depleted cells caused proton entry (as determined by pH measurements). Thus there is excellent support for the sugar–H^+ cotransport hypothesis for this system.

Evidence favoring a 1 : 1 stoichiometry of protons and sugar comes from studies comparing the steady-state protonmotive force and sugar accumulation ratio. If there were obligatory coupling between protons and sugar and a 1 : 1 stoichiometry, then the chemical potential difference for sugar ($\Delta\mu_{TMG}$) should equal the electrochemical potential for protons ($\Delta\overline{\mu}_{H^+}$). Experimentally determined values for $\Delta\mu_{TMG}$ and $\Delta\overline{\mu}_{H^+}$ agree and thus support the 1 : 1 stoichiometry in this cell (Kashket and Wilson, 1974; Wilson et al., 1978). Such data are sufficiently reproducible that sugar accumulation has been used to calculate (indirectly) the protonmotive force in this cell (Kashket and Barker, 1977).

4. L-ARABINOSE TRANSPORT

Although L-arabinose has fewer pathways than galactose for entry into E. coli, growth on arabinose causes induction of two transport systems, one of which requires a binding protein whereas the other involves H^+–sugar transport.

Transport of L-arabinose into E. coli B/r was described by Novotny and Englesberg in 1966. Induction was by L-arabinose alone, and energy-dependent accumulation of the sugar could be obtained in mutant cells lacking L-arabinose isomerase. Inhibition of this accumulation was effected by D-xylose, D-fucose, and D-galactose, suggesting that these sugars might be substrates for the system. Evidence for more than one transport pathway for arabinose in E. coli K_{12} was presented by Schleif (1969), who reported that two separate

K_m values for sugar uptake were obtained (5×10^{-5} M and 3×10^{-6} M). A periplasmic binding protein was found in induced cells of both B/r and K_{12} strains of *E. coli* (Hogg and Englesberg, 1969; Schleif, 1969). The K_m for the binding of arabinose and the patterns of inhibition agree with involvement of the binding protein with the high-affinity AraF system (Brown and Hogg, 1972). Both AraF and the lower-affinity transport system (AraE) are controlled by the regulatory gene *AraC* on the arabinose operon.

Henderson (1974) showed that addition of L-arabinose to anaerobic cells of *E. coli*, constitutive for L-arabinose transport, caused effective proton uptake as measured by alkalinization of the suspending medium. D-Fucose

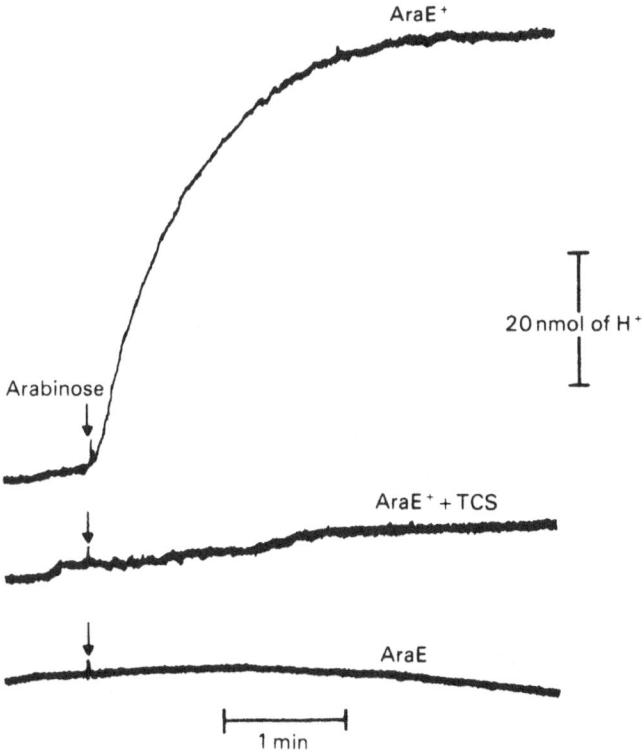

FIGURE 8. Arabinose-induced proton entry in AraE$^+$ strain. Strain SB5314 (AraE$^+$) or SB5313 (AraE$^-$) was grown on glycerol plus arabinose (as inducer). Washed anaerobic cells were incubated in 150 mM KCl–2 mM glycylglycine at pH 6.5 and the pH was monitored continuously. An upward deflection represents an alkaline pH change. TCS was added 10 min before arabinose in the second experiment. (From Daruwalla *et al.*, 1981.)

gave even greater proton movement than L-arabinose, whereas galactose had about 20% of the activity of arabinose and IPTG was inactive. Although these experiments established the fact that there was proton translocation with that of sugar, they did not distinguish between the two systems present for L-arabinose transport under these conditions.

Later work in the same laboratory (Daruwalla *et al.*, 1981) resolved this question. Strains of *E. coli* lacking either AraE or AraF were tested, as well as strains lacking both properties or retaining both. Only when AraE was present (regardless of whether or not AraF was retained) was H$^+$ uptake observed on the addition of L-arabinose to anaerobic resting cell suspensions. Figure 8 shows that the addition of arabinose to anaerobic cells of an AraE$^+$

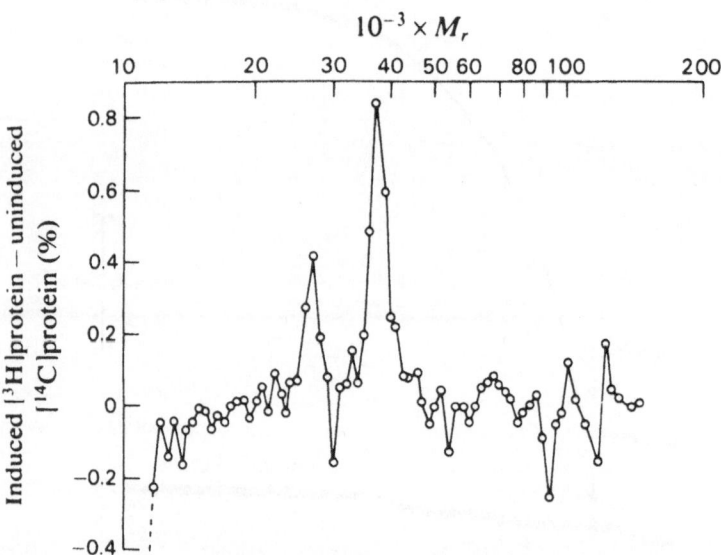

FIGURE 9. Dual-isotope labeling of strain SB5314. An arabinose-induced and an uninduced culture of SB5314 were grown in parallel on glycerol. After growth for one generation, 100 μCi of ^{3}H-labeled amino acid mixture was added to the induced culture and 50 μCi of ^{14}C-labeled amino acid mixture was added to the uninduced culture; growth was continued for another generation. The bacteria from the two cultures were harvested and pooled, and a single membrane preparation was made. The membranes were incubated for 15 min at 37°C in SDS dissolving buffer and two 50-μl samples were analyzed by SDS–polyacrylamide (15%) gel electrophoresis. After drying, each track was cut into 1-mm sections and incubated with 0.4 ml of H$_2$O$_2$ at 60°C for 12 hr before liquid scintillation counting. The difference between the percentages of radio-isotope in each section is plotted against the apparent relative molecular mass. The peak at 36,000 daltons represents the AraE gene product. (From Macpherson *et al.*, 1981.)

strain caused proton entry which was blocked by the proton ionophore tetrachlorosalicylanilide (TCS). In an AraE⁻ strain a similar addition of arabinose gave no proton movement. D-Fucose also was a good substrate for this system. In these experiments the cells had been induced with arabinose during growth; uninduced cells showed no pH change on the addition of the sugars.

The AraE transport protein of *E. coli* has been identified as a protein of apparent subunit relative molecular mass (M_r) 36,000–37,000 (Macpherson *et al.*, 1981). The dual-isotope labeling technique was used to distinguish between proteins produced by the arabinose-induced cell and those present in the uninduced cell (Figure 9). This method involved growth of one culture in the presence of ^{3}H-labeled amino acids and the other in the presence of ^{14}C-labeled amino acids. The two cultures (induced and uninduced) were pooled and the membrane proteins prepared. Analysis by two-dimensional gel electrophoresis showed that the protein labeled in the induced culture had two peaks at M_r 36,000 and 27,000 not seen in the uninduced cells (Figure 9). In an AraE deletion mutant which lacked H^+–arabinose cotransport, the protein at M_r 36,000 was no longer present. Both of these characteristics were restored when AraE$^+$ was transduced by phage into the deletion strain.

5. D-XYLOSE TRANSPORT

A transport system in *E. coli* induced by D-xylose has been shown to link proton uptake with the movement of this sugar (Lam *et al.*, 1980). Uncoupling agents inhibit the transport whereas arsenate and fluoride do not. That this system is distinct from previously described proton–sugar symport systems was shown by tests of inducer and substrate specificities. Thus growth in the presence of IPTG (inducer of LacY) or with D-fucose (inducer of GalP) failed to produce cells capable of proton–xylose symport. Although growth in arabinose (inducer of AraE) yielded cells that showed a very slight proton uptake on the addition of xylose, the value was only 10% of that obtained with arabinose, and when arabinose was tested as substrate of xylose-induced cells it was found to be inactive, as were fucose and galactose. Additional evidence for a distinct xylose–proton carrier was obtained by the mapping of a transport-negative mutant at minute 79 that differs from the location of the known carriers. Thus a distinct H^+–xylose transport protein appears to be present in induced cells of *E. coli*.

6. MELIBIOSE TRANSPORT

Although many bacterial cotransport systems use H^+ as the coupling ion, there are also certain examples of Na^+–substrate cotransport. The melibiose transport carrier of *E. coli* is a representative of this category. A remarkable feature of this carrier is the different cation requirement for different substrates (see Section 6.2).

In 1965 Prestidge and Pardee discovered that a distinct melibiose transport system could be induced in *E. coli* strains B and K_{12}. The melibiose carrier was found in lactose-deleted K_{12} strains and could be produced by growth in the presence of galactinol, which did not induce the lactose operon. In addition Prestidge and Pardee (1965) noted the interesting and most unusual fact that the melibiose carrier was temperature sensitive. *E. coli* K_{12} grown at 37° did not possess the transport system, in contrast to cells grown at 25°, in which active transport was present. However, in another strain of *E. coli* (strain B), in *Salmonella typhimurium*, and in *Klebsiella aerogenes* the carrier is stable at 37°.

TABLE IV.   Substrate and Inducer Specificity for the
Melibiose Carrier in *E. coli* and *Klebsiella aerogenes*[a]

	Substrate		Inducer	
Sugar	*E. coli*	*K. aerogenes*	*E. coli*	*K. aerogenes*
Melibiose	+	+	+	+
Lactose	−	+	−	−
Melibiitol	+		+	−
TMG	+	+	−	−
IPTG			+	−
Galactose	+		+	+
D-Fucose	−		−	−
Raffinose	+	+	+	+
α-PNPG	+	+		

[a] Data for *E. coli* from Pardee (1957) and Rotman *et al.* (1968). Data for *K. aerogenes* from Reeve and Braithwaite (1973) and T.H. Wilson and D.M. Wilson (unpublished observations).

6.1. Substrate and Inducer Specificities

Although the natural substrates for the melibiose carrier are α-galactosides, a variety of β-galactosides are also recognized. One of these is TMG, which has been used so extensively to study lactose transport. When induced cells are incubated with this sugar, accumulation within the cell occurs to concentrations as high as 200 times that in the incubation medium. A summary of the compounds tested as substrates and inducers is given in Table IV.

Several α-galactosides can induce the melibiose operon in *E. coli* without inducing the lactose system. Among these inducers are galactinol, melibiitol, and α-phenylgalactoside. Other sugars, including melibiose, induce both the melibiose as well as the lactose operons.

6.2. Cation Requirements

Stock and Roseman (1971) reported that Na^+ or Li^+ caused stimulation of TMG uptake in melibiose-induced *S. typhimurium*. This fact and the further observation that TMG stimulated Na^+ uptake provided strong evidence for Na^+–TMG cotransport in these cells. Confirmation of this work and further studies in this area were reported in both *S. typhimurium* (Tokuda and Kaback, 1977, 1978; Silva and Dobrogosz, 1978; Van Thienen *et al.*, 1978) and *E. coli* (Tsuchiya *et al.*, 1977; Lopilato *et al.*, 1978; Tsuchiya and Wilson, 1978; Wilson *et al.*, 1980; Tsuchiya *et al.*, 1980; Ottina *et al.*, 1980; Tanaka *et al.*, 1980; Cohn and Kaback, 1980).

TABLE V. Effects of Cations on Sugar Transport in the Melibiose System of *E. coli* K_{12}[a]

Sugar	Na^+	Li^+	H^+
Melibiose	Stimulate	Inhibit	Stimulate
α-PNPG	Inhibit	Inhibit	
Methyl-α-galactoside	Stimulate	Stimulate	Stimulate
TMG	Stimulate	Stimulate	No effect
Methyl-β-galactoside	Stimulate		No effect
Thiodigalactoside	Stimulate		No effect
β-PNPG	Stimulate	No effect	
Raffinose	Stimulate	No effect	No effect
Galactose	Stimulate		

[a] From Tsuchiya and Wilson (1978).

A striking property of the melibiose transport carrier is the remarkable variation in the cation requirement for the transport of different substrates (Table V). While most of the substrates can use Na^+ for cotransport, a few can also use Li^+ or H^+. The natural substrate melibiose may enter the cell on the carrier in cotransport with either Na^+ or H^+. Figure 10 shows that, when melibiose is added to anaerobic cells in the absence of Na^+, alkalinization of the medium is observed (Tsuchiya and Wilson, 1978). When the experiment is repeated in the presence of Na^+, acidification of the medium is observed. α-Methylgalactoside uptake in the absence of Na^+ is associated with proton entry into *E. coli*, while no such H^+ cotransport was observed with TMG, TDG, or β-methylgalactoside (Tsuchiya and Wilson, 1978). Evidence has been presented that melibiose–H^+ cotransport does not take place via the galactose (GalP) or the L-arabinose (AraE) transport system (Wilson *et al.*, 1982).

The melibiose carrier in *Klebsiella* does not show the same cation requirement as that seen in *E. coli* and *Salmonella*. Na^+ and Li^+ have no effect on the transport of either TMG or α-PNPG in *Klebsiella*, and Li^+ has no effect on the growth of this organism on melibiose. A LacY⁻ mutant of *Klebsiella* grown on melibiose showed an alkalinization of the incubation medium on the addition of TMG (D.M. Wilson and T.H. Wilson, unpublished observations), suggesting TMG–H^+ cotransport.

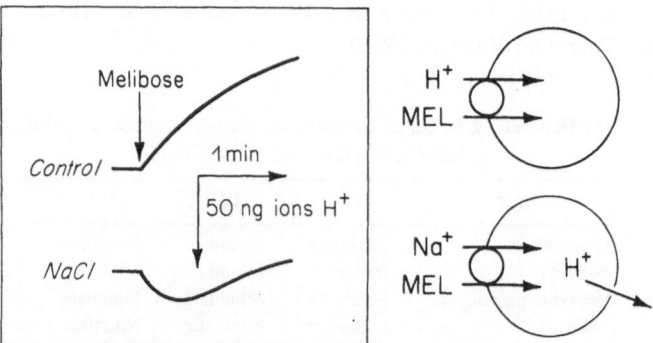

FIGURE 10. Proton movement induced by the addition of melibiose. Cells of RA11 (Lopilato *et al.*, 1978) were incubated anaerobically in the presence of 90 mM KCl and 30 mM KSCN with or without 10 mM NaCl. After preincubation for 45 min at room temperature, 20 μl of anaerobic 0.5 M melibiose was added to 2 ml of cell suspension and pH changes monitored. An upward deflection represents a rise in the pH of the medium. (From Wilson *et al.*, 1980.)

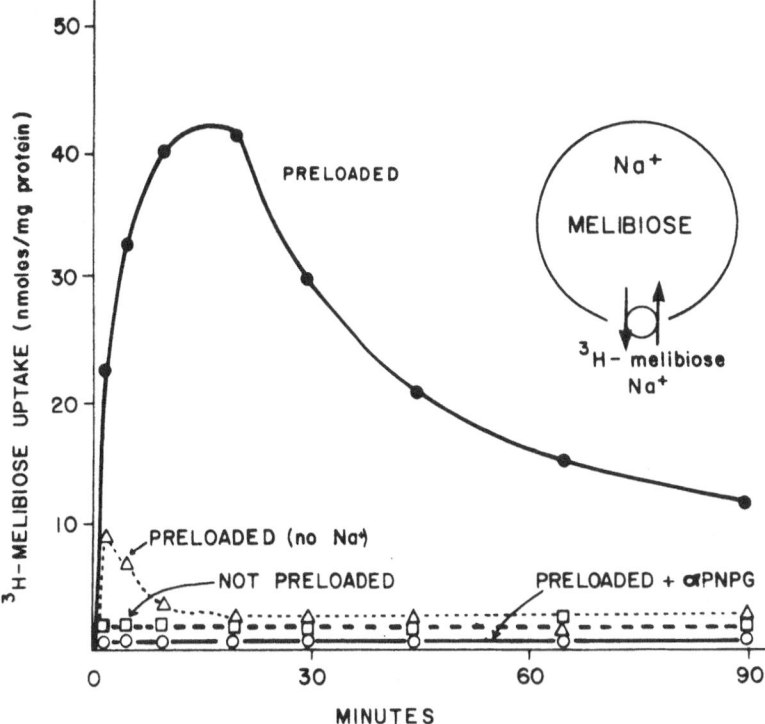

FIGURE 11. Counterflow of melibiose in reconstituted proteoliposomes. Reconstitution was carried out with an octylglucoside extract of RA11/pLC 25–33, a melibiose plasmid-containing strain. In three of the experiments proteoliposomes were preloaded with melibiose (20 mM), KPO₄ (40 mM), and NaCl (10 mM). Proteoliposomes were exposed to [³H]melibiose in all cases. In one case (△---△) proteoliposomes were preloaded with melibiose but no Na⁺ was present either inside or outside; (□--□) no sugar preloading but Na⁺ (10 mM) present both inside and outside; (○——○) αPNPG added externally as an inhibitor of transport. (From Tsuchiya et al., 1982.)

6.3. Reconstitution of the Melibiose Carrier

Tsuchiya et al. (1982) have reported the extraction of E. coli melibiose carrier from the membrane with octylglucoside and the reconstitution of carrier activity in proteoliposomes. Figure 11 shows the transport of [³H]melibiose by proteoliposomes preloaded with 20 mM nonradioactive melibiose plus 10 mM Na⁺ and diluted into medium containing 10 mM Na⁺ and [³H]melibiose. In this counterflow experiment the concentration of radioactive sugar in the proteoliposomes rose for 15 min and then declined. In the absence of preloaded sugar, melibiose uptake was very low. In this case the disaccharide presumably entered those liposomes containing carrier molecules until the concentrations of sugar inside and outside equilibrated. If this inference is correct the accumulation of radioactivity at 15 min in the preloaded proteoliposomes can be calculated to be 20-fold that in the medium. Preloading with sugar but without Na^+ and dilution into Na^+-free medium resulted in a small accumulation followed by a fall to levels found in the experiment with no preloading. In this case it was assumed that the melibiose was being transported associated with H^+ rather than Na^+. The inhibitor α-PNPG blocked entry of the radioactive sugar. It was also possible to demonstrate membrane-potential-driven melibiose accumulation in the presence of Na^+. In addition an Na^+ gradient (100 mM outside, 0 mM inside the liposome) also resulted in disaccharide accumulation. Thus the ion specificity found in intact cells could be reproduced in the reconstituted system.

6.4. Mutants

Tsuchiya et al. (1982) found that 10 mM LiCl strongly inhibited the growth of E. coli on minimal medium with melibiose as the sole carbon source. It has been possible to isolate a series of Li^+-resistant mutants with a variety of properties (Tsuchiya et al., 1978; Wilson et al., 1982; Niiya et al., 1982). One interesting mutant requires Na^+ or Li^+ for growth, in contrast to parental strains that grow well in the absence of these ions (Niiya et al., 1982). Unlike the parent it fails to show proton uptake when melibiose is added to anaerobic cells. Thus, the cell fails to show proton–melibiose transport. Apparently a mutation has occurred at the cation recognition site which allows it to use Na^+ or Li^+ for cotransport (but it cannot use H^+). Another Li^+-resistant mutant has been isolated that not only grows well in the absence of Na^+ (as does the parent) but exhibits less of a stimulatory effect of Na^+ on the transport of sugars in washed cells (Wilson et al., 1982). This mutant has an altered cation recognition site that has partially lost its Na^+ stimulation but retains [H^+]melibiose cotransport.

These mutants are of interest in relation to the hypothesis (Wilson and Maloney, 1976) that during evolution the earliest cells showed proton gradients and proton–substrate cotransport, while at a later stage Na^+ gradients appeared and Na^+–substrate cotransport predominated (as in animal cells).

According to this view there must have been an intermediate stage of some carrier proteins in which recognition of both cations occurred. It is of interest to speculate that the melibiose carrier of *E. coli* might represent a descendant of one of the "transition" forms of cation–substrate cotransport. It might be expected that mutants could be found that show a marked preference for either H^+ or Na^+. It remains to be seen from structural studies of these mutants whether relatively simple amino acid substitutions in the region of the cation recognition site might greatly alter the cation used for cotransport.

Another type of mutant has been isolated that affects several Na^+-dependent transport systems. Krulwich *et al.* (1979) isolated mutants of *Bacillus alkalophilis* that show defects in the Na^+–H^+ antiporter and several Na^+–substrate cotransport systems (Guffanti *et al.*, 1981). A similar mutant has been isolated by Zilberstein *et al.* (1982) in *E. coli;* it shows a pleotropic defect in the Na^+–H^+ antiporter, Na^+–melibiose cotransport, and Na^+–glutamic acid cotransport. It is postulated by both groups of investigators that there is an Na^+-coupling protein needed for each of the Na^+-dependent translocation processes. Thus, for example, melibiose transport requires both the product of the *MelB* gene and the Na^+-coupling protein.

In summary the melibiose transport system shows a more complex cation requirement than is found in other cotransport systems. The carrier utilizes H^+ for α-PNPG, Na^+ or Li^+ for TMG, and H^+ or Na^+ for melibiose. The strong inhibition of growth by Li^+ in the presence of melibiose has permitted the isolation of a variety of Li^+-resistant mutants. Among such mutants was found one which has lost its ability to cotransport with protons, while retaining its ability to recognize Na^+ or Li^+; a second mutant has partially lost its ability to recognize Na^+ while retaining its use of the protons for cotransport.

7. SUMMARY

The cation–substrate cotransport system is a membrane-translocating device that utilizes the existing cation gradients as a source of energy for the cellular accumulation of substrates in the external environment. In the extensively studied system of lactose–proton cotransport in *E. coli*, genetic and biochemical investigations have revealed the complete DNA sequence of the *lacY* gene. The amino acid sequence inferred from this data indicates that the protein is extremely hydrophobic. This protein has been purified by several techniques and reconstituted into proteoliposomes. It is believed that a single polypeptide is responsible for the biological activity of membrane translocation.

Recently considerable progress has been made in the study of other transport systems, particularly those for galactose, arabinose, and melibiose in *E. coli*. They have been characterized physiologically and the first reports of biochemical characterization have appeared. The arabinose carrier has been identified in two-dimensional gels and has been specifically labeled with NEM. The melibiose carrier has been extracted from the membrane and reconstituted in proteoliposomes. It is anticipated that in the near future biochemical and genetic studies will provide information on the anatomical arrangement of these proteins within the membrane. Furthermore it is likely that the DNA sequence of many of these carrier proteins will be available shortly, and comparison with the lactose carrier will be possible. The kinetic data combined with structural information should help elucidate the biochemical mechanism of transport of sugars with various cations.

REFERENCES

Beyreuther, K., Bieseler, B., Ehring, R., and Müller-Hill, B., 1981, Identification of internal residues of lactose permease of *Escherichia coli* by radiolabel sequencing of peptide mixtures, in: *Methods in Protein Sequence Analysis* (M. Elzinga, ed.), Humana Press, Clifton, New Jersey, pp. 139–148.

Beyreuther, K., and Ehring, R., 1980, Microsequence analysis of lactose permease of *Escherichia coli*, in: *Methods in Peptide and Protein Sequence Analysis* (C. Birr, ed.), Elsevier Biomedical Press, Amsterdam, pp. 199–212.

Beyreuther, K., Bieseler, B., Ehring, R., Griesser, H.-W., Mieschendahl, M., Müller-Hill, B., and Triesch, I., 1980, Active transport in microorganisms, *Biochem. Soc. Trans.* **8:**675–676.

Bihler, I., and Crane, R. K., 1962, Studies on the mechanism of intestinal absorption of sugars. V. The influence of several cations and anions on the active transport of sugars *in vitro* by various preparations of hamster small intestine, *Biochim. Biophys. Acta* **59:**78–93.

Booth, I. R., and Hamilton, W. A., 1980, Quantitative analysis of proton-linked transport systems: β-Galactoside exit in *Escherichia coli*, *Biochem. J.* **188:**467–473.

Brown, C. E., and Hogg, R. W., 1972, A second transport system for L-arabinose in *Escherichia coli* B/r controlled by the *araC* gene, *J. Bacteriol.* **11:**606–613.

Büchel, D. E., Gronenborn, B., and Müller-Hill, B., 1980, Sequence of the lactose permease gene, *Nature* **283:**541–545.

Carter, J. R., Fox, C. F., and Kennedy, E. P., 1968, Interaction of sugars with the membrane protein component of the lactose transprt system of *Escherichia coli*, *Proc. Natl. Acad. Sci. USA* **60:**725–732.

Christensen, H. N., and Riggs, T. R., 1952, Concentrative uptake of amino acids by the Ehrlich mouse ascites carcinoma cell, *J. Biol. Chem.* **194:**57–68.

Cohen, G. N., and Monod, J., 1957, Bacterial permeases, *Bacteriol Rev.* **21:**169–194.

Cohn, D. E., and Kaback, J. R., 1980, Mechanism of the melibiose porter in membrane vesicles of *Escherichia coli*, *Biochemistry* **19:**4237–4243.

Cohn, D. E., Kaczorowski, G. J., and Kaback, H. R., 1981, Effect of the proton electrochemical gradient on maleimide inactivation of active transport in *Escherichia coli* membrane vesicles, *Biochemistry* **20:**3308–3313.

Crane, R. K., 1962, Hypothesis for mechanism of intestinal active transport of sugars, *Fed. Proc.* **21:**891–895.

Crane, R. K., 1965, Na$^+$-dependent transport in the intestine and other animal cells, *Fed. Proc.* **24:**1000–1006.

Crane, R. K., 1977, The gradient hypothesis and other models of carrier-mediated active transport, *Rev. Physiol. Biochem. Pharmacol.* **78:**99–159.

Csaky, T. Z., and Thale, M., 1960, Effect of ionic environment on intestinal sugar transport, *J. Physiol. (London)* **151:**59–65.

Daruwalla, K. R., Paxton, A. T., and Henderson, P. J. F., 1981, Energization of the transport systems for arabinose and comparison with galactose transport in *Escherichia coli*, *Biochem. J.* **200:**611–627.

Desai, P. D., and Goldner, M., 1969, Effect of low pH on thiomethyl-β-D-galactoside uptake by *Streptococcus lactis*, *J. Bacteriol.* **100:**1415–1416.

Eddy, A. A., 1978, Proton-dependent solute transport in microorganisms, in: *Current Topics in Membrane Transport*, Volume 10 (F. Bronner and A. Kleinzeller, eds.), Academic Press, London, pp. 279–360.

Eddy, A. A., and Nowacki, J. A., 1971, Stoichiometrical proton and potassium ion movements accompanying the absorption of amino acids by the yeast *Saccharomyces carlsbergensis*, *Biochem. J.* **122:**701–711.

Ehring, R., Beyreuther, K., Wright, J. K., and Overath, P., 1980, *In vitro* and *in vivo* products of *E. coli* lactose permease gene are identical, *Nature* **283:**537–540.

Flagg, J. L., and Wilson, T. H., 1976, *Lac*Y mutant of *Escherichia coli* with altered physiology of lactose induction; *J. Bacteriol.* **128:**701–707.

Fox, C. F., and Kennedy, E. P., 1965, Specific labeling and partial purification of the M protein, a component of the β-galactoside transport system of *Escherichia coli*, *Proc. Natl. Acad. Sci. USA* **54:**891–899.

Fox, C. F., Carter, J. R., and Kennedy, E. P., 1967, Genetic control of the membrane protein component of the lactose transport system of *Escherichia coli*, *Proc. Natl. Acad. Sci. USA* **57:**698–705.

Fraenkel, D. G., Falcoz-Kelley, F., and Horecker, B. L., 1964, The utilization of glucose 6-phosphate by glucokinaseless and wild-type strains of *Escherichia coli*, *Proc. Natl. Acad. Sci. USA* **52:**1207–1213.

Fried, V. A., 1977, A novel mutant of the *lac* transport system of *E. coli*, *J. Mol. Biol.* **114:**477–490.

Fried, V. A., 1981, Membrane biogenesis: Evidence that a soluble chimeric polypeptide can serve as a precursor of a mutant *lac* permease in *Escherichia coli*, *J. Biol. Chem.* **256:**244–252.

Ganesan, A. K., and Rotman, B., 1965, Transport systems for galactose and galactosides in *Escherichia coli*. I. Genetic determination and regulation of the methyl-galactoside permease, *J. Mol. Biol.* **16:**42–50.

Guffanti, A. A., Cohn, D. E., Kaback, H. R., and Krulwich, T. A., 1981, Relationship between the Na$^+$/H$^+$ antiporter and Na$^+$/substrate symport in *Bacillus alcalophilus*, *Proc. Natl. Acad. Sci. USA* **78:**1481–1484.

Hamilton, W. A., 1975, Energy coupling in microbial transport, *Adv. Microb. Physiol.* **12**:1–53.

Hamilton, W. A., 1977, Energy coupling in substrate and group translocation, in: *Microbiol Energetics* (B. A. Haddock and W. A. Hamilton, eds.), Society for General Microbiology, Symposium 27, Cambridge University Press, Cambridge, pp. 185–216.

Harold, F. M., 1978, Vectorial metabolism, in: *The Bacteria* (I. C. Gunsalus, ed.), Volume VI, *Bacterial Diversity* (L. N. Ornstron and J. R. Sokatch, eds), Academic Press, New York, pp. 465–513.

Hendersen, P. J. F., 1974, Application of the chemiosmotic theory to the transport of lactose, D-galactose and L-arabinose by *Escherichia coli*, in: *Comparative Biochemistry and Physiology of Transport* (L. Bolis, K. Bloch, S. E. Luria, and F. Lynen, eds.), North-Holland, Amsterdam, pp. 409–424.

Henderson, P. J. F., and Giddens, R. A., 1977, 2-Deoxy-D-galactose, a substrate for the galactose-transport system of *Escherichia coli*, *Biochem. J.* **168**:15–22.

Henderson, P. J. F., Giddens, R. A., and Jones-Mortimer, M. C., 1977, Transport of galactose, glucose and their molecular analogs by *Escherichia coli* K12, *Biochem. J.* **162**:309–320.

Herzenberg, L. A., 1959, Studies of the induction of β-galactosidase in a cryptic strain in *Escherichia coli*, *Biochim Biophys. Acta* **31**:525–538.

Hirata, H., Altendorf, K., and Harold, F. M., 1974, Energy coupling in membrane vesicles of *Escherichia coli*, *J. Biol. Chem.* **249**:2939–2945.

Hobson, A. C., Gho, D., and Müller-Hill, B., 1977, Isolation, genetic analysis, and characterization of *Escherichia coli* mutants with defects in the *lac*Y gene, *J. Bacteriol.* **131**:830–838.

Höfer, M., Misra, P. C., and Dahle, P., 1974, On the uniformity of the energy-dependent (nonphosphorylating) monosaccharide transport across the cell membrane of eukaryotes, in: *Proceedings of the 4th International Symposium on Yeasts*, Vienna, pp. 287–288.

Hogg, R. W., and Englesberg, E., 1969, L-Arabinose binding protein from *Escherichia coli* B/r, *J. Bacteriol.* **100**:423–432.

Hong, J.-S., 1977, An *ecf* mutation in *Escherichia coli* pleiotropically affecting energy coupling in active transport but not generation or maintenance of membrane potential, *J. Biol. Chem.* **252**:8582–8588.

Horecker, B. L., Thomas, J., and Monod, J., 1960a, Galactose transport in *Escherichia coli*. I. General properties as studied in a galactokinaseless mutant, *J. Biol. Chem.* **235**:1580–1585.

Horecker, B. L., Thomas, J., and Monod, J., 1960b, Galactose transport in *Escherichia coli*. II. Characteristics of the exit process, *J. Biol. Chem.* **235**:1586–1590.

Jones, T. H. D., and Kennedy, E. P., 1969, Characterization of the membrane protein of the lactose transport system of *Escherichia coli*, *J. Biol. Chem.* **244**:5981–5987.

Kaczorowski, G. J., and Kaback, H. R., 1979a, Mechanism of lactose translocation in membrane vesicles from *Escherichia coli*. 1. Effect of pH on efflux, exchange, and counterflow, *Biochemistry* **18**:3691–3697.

Kaczorowski, G. J., and Kaback, H. R., 1979b, Mechanism of lactose translocation in membrane vesicles from *Escherichia coli*. 2. Effect of imposed $\Delta\psi$, ΔpH, and $\Delta\bar{\mu}_{H^+}$, *Biochemistry* **18**:3697–3704.

Kashket, E. R., and Barker, S. L., 1977, Effect of potassium ions on the electrical and pH gradients across the membrane of *Streptococcus lactis* cells, *J. Bacteriol.* **130**:1017–1023.

Kashket, E. R., and Wilson, T. H., 1972a, Role of metabolic energy in the transport of β-galactosides by *Streptococcus lactis*, *J. Bacteriol.* **109**:784–789.

Kashket, E. R., and Wilson, T. H., 1972b, Galactoside accumulation associated with ion movements in *Streptococus lactis*, *Biochim. Biophys. Acta* **49**:615–620.

Kashket, E. R., and Wilson, T. H., 1974, Protonmotive force in fermenting *Streptococcus lactis* 7962 in relation to sugar accumulation, *Biochem. Biophys. Res. Commun.* **59**:879–886.

Kennedy, E. P., 1970, The lactose permease system of *Escherichia coli*, in: *The Lactose Operon* (J. R. Beckwith and D. Zipser, eds.), Cold Spring Harbor Laboratory, Cold Spring Harbor, New York, pp. 49–92.

Kennedy, E. P., Rumley, M. K., and Armstrong, J. B., 1974, Direct measurement of the binding of labelled sugars to the lactose permease M protein, *J. Biol. Chem.* **249:**33–37.

Kepes, A., 1960, Etudes cinétiques sur la galactoside-perméase d'*Escherichia coli*, *Biochim. Biophys. Acta* **40:**70–84.

Koch, A. L., 1964, The role of permease in transport, *Biochim. Biophys. Acta* **79:**177–200.

Komor, E., 1973, Proton-coupled hexose transport in *Chlorella vulgaris*, *FEBS Lett.* **38:**16–18.

Kornberg, H. L., 1976, Genetics in the study of carbohydrate transport by bacteria (6th Griffith Memorial Lecture), *J. Gen. Microbiol.* **96:**1–16.

Kornberg, H. L., and Riordan, C. L., 1976, Uptake of galactose into *Escherichia coli* by facilitated diffusion, *J. Gen. Microbiol.* **94:**75–89.

Krulwich, T. A., Mandel, K. G., Bornstein, R. F., and Guffanti, A. A., 1979, A non-alkalophilic mutant of *Bacillus alcalophilus* lacks the Na^+/H^+ antiporter, *Biochem. Biophys. Res. Commun.* **91:**58–62.

Kusch, M., and Wilson, T. H., 1973, Defective lactose utilization by a mutant of *Escherichia coli* energy-uncoupled for lactose transport: The advantages of active transport versus facilitated diffusion, *Biochim. Biophys. Acta* **311:**109–122.

Lam, V. M. S., Daruwalla, K. R., Henderson, P. J. F., and Jones-Mortimer, M. C., 1980, Proton-linked D-xylose transport in *Escherichia coli*, *J. Bacteriol.* **143:**396–402.

Lancaster, J. R., Jr., Hill, R. J., and Struve, W. G., 1975, The characterization of energized and partially de-energized respiration-independent β-galactoside transport into *Escherichia coli*, *Biochim. Biophys. Acta* **401:**285–298.

Langridge, J., 1974, Characterization and intragenic position of mutations in the gene for galactoside permease of *Escherichia coli*, *Aust. J. Biol. Sci.* **27:**331–340.

LeQuoc, D., LeQuoc, K., and Gaudemer, Y., 1977, Influence of the energetic state of rat liver mitochondria on the sensitivity of the phosphate carrier towards SH reagents, *Biochim. Biophys. Acta* **462:**131–140.

Lopilato, J., Tsuchiya, T., and Wilson, T. H., 1978, Role of Na^+ and Li^+ in thiomethylgalactoside transport by the melibiose transport system of *Escherichia coli*, *J. Bacteriol.* **134:**147–156.

Macpherson, A. J. S., Jones-Mortimer, M. C., and Henderson, P. J. F., 1981, Identification of the AraE transport protein of *Escherichia coli*, *Biochem. J.* **196:**269–283.

Maloney, P. C., and Wilson, T. H., 1978, Metabolic control of lactose entry in *Escherichia coli*, *Biochim. Biophys. Acta* **511:**487–498.

Manno, J. A., and Schachter, D., 1970, Energy-coupled influx of thiomethylgalactoside into *Escherichia coli*, *J. Biol. Chem.* **245:**1217–1223.

Mieschendahl, M., Büchel, D., Bocklage, H., and Müller-Hill, B., 1981, Mutations in the *lac*Y gene of *Escherichia coli* define functional organization of lactose permease, *Proc. Natl. Acad. Sci. USA* **78:**7652–7656.

Mitchell, P., 1963, Molecule, group and electron translocation through natural membranes, *Biochem. Soc. Symp.* **22:**142–168.

Mitchell, P., 1968, *Chemiosmotic Coupling and Energy Transduction*, pp. 1–111, Glynn Research Laboratories, Bodmin, Cornwall, England.

Mitchell, P., 1976, The 9th CIBA Medal Lecture. Vectorial chemistry and the molecular mechanics of chemiosmotic coupling: Power transmission by proticity, *Biochem. Soc. Trans.* **4:**399–430.

Mitchell, P., 1977, Epilogue: From energetic abstraction to biochemical mechanism, *Symp. Soc. Gen. Microbiol.* **27:**383–423.

Mitchell, P., and Moyle, J., 1974, The mechanisms of proton translocation in reversible proton-translocating adenosine triphosphatases, in: *Membrane Adenosine Triphosphatases and Transport Processes* (J. R. Bronk, ed.), *Biochem. Soc. Spec. Publ.* **4:**91–111.

Newman, M. J., and Wilson, T. H., 1980, Solubilization and reconstitution of the lactose transport system from *Escherichia coli, J. Biol. Chem.* **255:**10583–10586.

Newman, M. J., Foster, D. L., Wilson, T. H., and Kaback, H. R., 1981, Purification and reconstitution of functional lactose carrier from *Escherichia coli, J. Biol. Chem.* **256:**11804–11808.

Niiya, S., Yamasaki, K., Wilson, T. H., and Tsuchiya, T., 1982, Altered coupling to melibiose transport in mutants of *Escherichia coli, J. Biol. Chem.* **257:**8902–8906.

Novotny, C. P., and Englesberg, E., 1966, The L-arabinose permease system in *Escherichia coli* B/r, *Biochim. Biophys. Acta* **117:**217–230.

Ottina, K., Lopilato, J., and Wilson, T. H., 1980, Membrane transport of *p*-nitrophenyl-α-galactoside by the melibiose carrier of *Escherichia coli, J. Membr. Biol.* **56:**169–175.

Padan, E., Patel, L., and Kaback, H. R., 1979a, Effect of diethylpyrocarbonate on lactose/proton symport in *Escherichia coli* membrane vesicles, *Proc. Natl. Acad. Sci USA* **76:**6221–6225.

Padan, E., Schuldiner, S., and Kaback, H. R., 1979b, Reconstitution of *lac* carrier function in cholate-extracted membranes from *Escherichia coli, Biochem. Biophys. Res. Commun.* **91:**854–861.

Page, M. G. P., and West, I. C., 1980, Kinetics of lactose transport into *Escherichia coli* in the presence and absence of a protonmotive force, *FEBS Lett.* **120:**187–191.

Pardee, A. B., 1957, An inducible mechanism for accumulation of melibiose in *Escherichia coli, J. Bacteriol.* **73:**376–385.

Plate, C. A., and Suit, J. L., 1981, The *eup* genetic locus of *Escherichia coli* and its role in H$^+$/solute symport, *J. Biol. Chem.* **256:**12974–12980.

Postma, P., 1976, Involvement of the phosphotransferase system in galactose transport in *Salmonella typhimurium, FEBS Lett.* **61:**49–53.

Prestidge, L. S., and Pardee, A. B., 1965, A second permease for methyl-thio-β-D-galactoside in *Escherichia coli, Biochim. Biophys. Acta* **100:**591–593.

Putzrath, R. M., and Wilson, T. H., 1979, Transport of alpha-*p*-nitrophenyl-galactoside by the lactose carrier of *Escherichia coli, J. Bacteriol.* **137:**1037–1039.

Racker, E., Violand, B., O'Neal, S., Alfonzo, M., and Telford, J., 1979, Reconstitution, a way of biochemical research; some new approaches to membrane-bound enzymes, *Arch. Biochem. Biophys.* **198:**470–477.

Reeve, E. C. R., and Braithwaite, J. A., 1973, The lactose system in *Klebsiella aerogenes* V9A. II. Galactose permeases which accumulate lactose or melibiose *Genet. Res.* **21:**273–285.

Rickenberg, H. V., Cohen, G. N., Butlin, G., and Monod, J., 1956, La galactoside-permease d'*Escherichia coli, Ann. Inst. Pasteur Paris* **91:**829–857.

Riklis, E., and Quastel, J. H., 1958, Effects of cations on sugar absorption by isolated surviving guinea pig intestine, *Can. J. Physiol.* **36:**347–362.

Robbie, J. P., and Wilson, T. H., 1969, Transmembrane effects of β-galactosides on thiomethyl-β-galactoside transport in *Escherichia coli, Biochim. Biophys. Acta* **173:**234–244.

Robertson, D. E., Kaczorowski, G. J., Garcia, M.-L., and Kaback, H. R., 1980, Active transport in membrane vesicles from *Escherichia coli:* The electrochemical proton gradient alters the distribution of the *lac* carrier between two different kinetic states, *Biochemistry* **19:**5692–5702.

Rosen, B. P., and Kashket, E. R., 1978, Energetics of active transport, in: *Bacterial Transport* (B. P. Rosen, ed.), Marcel Dekker, Basel, pp. 559–620.

Rotman, B., and Radojkovic, J., 1964, Galactose transport in *Escherichia coli:* The mechanism underlying the retention of intracellular galactose, *J. Biol. Chem.* **239:**3153–3156.

Rotman, B., Ganesan, A. K., Guzman, R., 1968, Transport sytems for galactose and galactosides in *Escherichia coli.* II. Substrate and inducer specificities, *J. Mol. Biol.* **36:**247–260.

Saier, M. H., Jr., Straud, H., Massman, L. S., Judice, J. J., Newman, M. J., and Feucht, B. U., 1978, Permease-specific mutations in *Salmonella typhimurium* and *Escherichia coli* that release the glycerol, maltose, melibiose, and lactose transport systems from regulation by the phosphoenolpyruvate:sugar phosphotransferase system, *J. Bacteriol.* **133:**1358–1367.

Sandermann, H., Jr., 1977, β-D-Galactoside transport in *Escherichia coli:* Substrate recognition, *Eur. J. Biochem.* **80:**507–515.

Schleif, R., 1969, An L-arabinose binding protein and arabinose permeation in *E. coli* (K12), *J. Mol. Biol.* **46:**185–196.

Schuldiner, S., and Kaback, H. R., 1977, Fluorescent galactosides as probes for the *lac* carrier protein, *Biochim. Biophys. Acta* **472:**399–418.

Schultz, S. G., and Curran, P. F., 1970, Coupled transport of sodium and organic solutes, *Physiol. Rev.* **50:**637–718.

Schwab, W. G. W., and Komor, E., 1978, A possible role of the membrane potential in proton-sugar cotransport of *Chlorella, FEBS Lett.* **87:**157–160.

Shuman, H. A., and Beckwith, J., 1979, *Escherichia coli* K-12 mutants that allow transport of maltose via the β-galactoside transport system, *J. Bacteriol.* **137:**365–373.

Silhavy, T. J., Ferenci, T., and Boos, W., 1978, Sugar transport systems in *Escherichia coli,* in: *Bacterial Transport* (B. P. Rosen, ed.), Marcel Dekker, New York, pp. 127–169.

Silva, D., and Dobrogosz, W. J., 1978, Proton efflux associated with melibiose permease activity in *Salmonella typhimurium, Biochem. Biophys. Res. Commun.* **81:**750–755.

Simoni, R. D., and Postma, P. W., 1975, The energetics of bacterial active transport, *Annu. Rev. Biochem.* **43:**523–554.

Sistrom, W. R., 1958, On the physical state of the intracellularly accumulated substrates of β-galactoside-permease in *Escherichia coli, Biochim. Biophys. Acta* **29:**579–587.

Slayman, C. L., and Slayman, C. W., 1974, Depolarization of the plasma membrane of *Neurospora* during active transport of glucose: Evidence for a proton-dependent cotransport system, *Proc. Natl. Acad. Sci. USA* **71:**1935–1939.

Stock, J., and Roseman, S., 1971, A sodium-dependent sugar co-transport system in bacteria, *Biochem. Biophys. Res. Commun.* **44:**132–138.

Tanaka, K., Niiya, S., and Tsuchiya, T., 1980, Melibiose transport in *Escherichia coli, J. Bacteriol.* **141:**1031–1036.

Teather, R. M., Müller-Hill, B., Abrutsch, U., Aichele, G., and Overath, P., 1978, Amplification of the lactose carrier protein in *Escherichia coli* using a plasmid vector, *Mol. Gen. Genet.* **159:**239–248.

Teather, R. M., Bramhall, J., Riede, I., Wright, J. K., Furst, M., Aichele, G., Wilhelm, U., and Overath, P., 1980, Lactose carrier protein of *Escherichia coli:* Structure and expression of plasmids carrying the Y gene of the *lac* operon, *Eur. J. Biochem.* **108:**223–231.

Thompson, J., 1980, Galactose transport systems in *Streptococcus lactis, J. Bacteriol.* **144:**683–691.

Tokuda, H., and Kaback, H. R., 1977, Sodium-dependent methyl 1-thio-beta-D-galactopyranoside transport in membrane vesicles isolated from *Salmonella typhimurium, Biochemistry* **16:**2130–2136.

Tokuda, H., and Kaback, H. R., 1978, Sodium-dependent binding of *p*-nitrophenyl-α-D-galactopyranoside to membrane vesicles isolated from *Salmonella typhimurium, Biochemistry* **17:**698–705.

Tsuchiya, T., and Wilson, T. H., 1978, Cation–sugar cotransport in the melibiose transport system of *Escherichia coli*, *Membr. Biochem.* **2**:63–79.

Tsuchiya, T., Raven, J., and Wilson, T. H., 1977, Co-transport of Na⁺ and methyl-beta-D-thiogalactopyranoside mediated by the melibiose transport system of *E. coli*, *Biochem. Biophys. Res. Commun.* **76**:26–31.

Tsuchiya, T., Lopilato, J., and Wilson, T. H., 1978, Effect of lithium ion on melibiose transport in *Escherichia coli*, *J. Membr. Biol.* **42**:45–59.

Tsuchiya, T., Takeda, K., and Wilson, T. H., 1980, H⁺-substrate cotransport by the melibiose membrane carrier in *Escherichia coli*, *Membr. Biochem.* **3**:131–146.

Tsuchiya, T., Ottina, K., Moriyama, Y., Newman, M. J., and Wilson, T. H., 1982, Solubilization and reconstitution of the melibiose carrier from a plasmid-carrying strain of *Escherichia coli*, *J. Biol. Chem.* **257**:50125–50128.

Van Thienen, G. M., Postma, P. W., and Van Dam, K., 1978, Na⁺-dependent methyl beta-thiogalactoside transport in *Salmonella typhimurium*, *Biochim. Biophys. Acta* **513**:395–400.

Villarejo, M., 1980, Evidence for two *lac* Y gene derived protein products in the *E. coli* membrane, *Biochem. Biophys. Res. Commun.* **93**:16–23.

Villarejo, M., and Ping, C., 1978, Localization of the lactose permease protein(s) in the *E. coli* envelope, *Biochem. Biophys. Res. Commun.* **83**:935–942.

West, I., 1970, Lactose transport coupled to proton movements in *Escherichia coli*, *Biochem. Biophys. Res. Commun.* **41**:655–661.

West, I., 1973, Proton movements coupled to the transport of β-galactosides into *Escherichia coli*, in: *Ion Transport in Plants* (W. P. Anderson, ed.), Academic Press, New York, pp. 237–250.

West, I., 1980, Energy coupling in secondary active transport, *Biochim. Biophys. Acta* **604**:91–126.

West, I. C., and Wilson, T. H., 1973, Galactoside transport dissociated from proton movement in mutants of *Escherichia coli*, *Biochem. Biophys. Res. Commun.* **50**:551–558.

Wilson, D. B., 1974, The regulation and properties of the galactose transport system in *Escherichia coli* K12, *J. Biol. Chem.* **249**:553–558.

Wilson, T. H., 1978, Lactose transport in *Escherichia coli*, in: *Physiology of Membrane Disorders* (T. E. Andreoli, J. F. Hoffman, and D. P. Fanestil, eds.), Plenum Press, New York, pp. 459–476.

Wilson, T. H., and Maloney, P. C., 1976, Speculations on the evolution of ion transport mechanisms, *Fed. Proc.* **35**:2174–2179.

Wilson, T. H., Kusch, M., and Kashket, E. R., 1970, a mutant in *Escherichia coli* energy-uncoupled for lactose transport; a defect in the lactose-operon, *Biochem. Biophys. Res. Commun.* **40**:1407–1414.

Wilson, T. H., Flagg, J. L., and Wilson, D. M., 1978, H⁺–sugar cotransport in bacteria, *Acta Physiol. Scand. Suppl.* pp. 389–397.

Wilson, T. H., Tsuchiya, T., Lopilato, J., and Ottina, K., 1980, The cation specificity for the melibiose transport system of *E. coli*, *Ann. N.Y. Acad. Sci.* **341**:465–472.

Wilson, T. H., Ottina, K., and Wilson, D. M., 1982, Melibiose transport in bacteria, in: *Membranes and Transport*, Volume 2 (A. N. Martonosi, ed.), Plenum Press, New York, pp. 33–39.

Winkler, H. H., and Wilson, T. H., 1966, The role of energy coupling in the transport of β-galactosides by *Escherichia coli*, *J. Biol. Chem.* **241**:2200–2211.

Wong, P. T. S., Kashket, E. R., and Wilson, T. H., 1970, Energy coupling in the lactose transport system of *Escherichia coli*, *Proc. Natl. Acad. Sci. USA* **65**:63–69.

Wright, J. K., Teather, R. M., and Overath, P., 1979, Lactose carrier protein of *Escherichia coli:* The riddle of the two binding sites, in: *Functional and Molecular Aspects of Bio-*

membrane Transport (E. M. Klingenberg, F. Palmieri, and E. Quagliariello, eds.), Elsevier/ North-Holland Biomedical Press, Amsterdam, pp. 239–248.

Wright, J. K., Riede, I., and Overath, P., 1981, Lactose carrier protein of *Escherichia coli:* Interaction with galactosides, *Biochemistry* **20:**6404–6415.

Wright, J. K., Schwarz, H., Straub, E., Bieseler, B., and Beyreuther, K., 1982, Lactose carrier protein of *Escherichia coli:* Reconstitution of galactoside binding and countertransport, *Eur. J. Biochem.* **124:**545–552.

Zilberstein, D., Ophir, I. J., Padan, E., and Schuldiner, S., 1982, Na$^+$ gradient-coupled porters of *Escherichia coli* share a common subunit, *J. Biol. Chem.* **257:**3692–3696.

THE STRUCTURE AND FUNCTION OF BAND 3

Ian G. Macara and Lewis C. Cantley

1. INTRODUCTION

Band 3 is a 95,000-dalton glycoprotein that spans the bilayer of the red cell plasma membrane and catalyzes the exchange of anions (Cl^- and HCO_3^- *in vivo*) across the membrane. It is one of only a very few proteins known unequivocally to be involved in anion transport (the others being the ADP/ATP exchanger and the phosphate transporter of the mitochondrial inner membrane), and it has been the most intensively studied of these proteins. The widespread interest in Band 3 is not surprising in view of its ready availability. Red blood cells are easily acquired and Band 3 constitutes about 25% of the total red cell membrane protein (Fairbanks *et al.*, 1971; Steck, 1974). The purity can be raised to about 70% by a single-step procedure to remove peripheral membrane proteins.

Because a number of excellent and comprehensive reviews on Band 3 and anion exchange have been published in recent years (Rothstein *et al.*,

Abbreviations used in text: Kd, kilodalton; BIDS, 4-benzamido-4'-isothiocyanostilbene-2,2'-disulfonate; DBDS, 4,4'-dibenzamidostilbene-2,2'-disulfonate; DIDS, 4,4'-isothiocyanostilbene-2,2'-disulfonate; DNDS, ˙4,4'-dinitrostilbene-2,2'-disulfonate; DNFB, dinitrofluorobenzene; GPDH, glyceraldehyde-3-phosphate dehydrogenase; H₂DBDS, 4,4'-dibenzamidodihydrostilbene-2,2'-disulfonate; H₂DIDS, 4,4'-diisothiocyano-2,2'-dihydrostilbene disulfonate; H₂(NBD)₂DS, 4,4'-di(4-nitrobenz-2-oxo-1,3-diazole)-2,2'-dihydrostilbene disulfonate; NAP-taurine, *N*-(4-azido-2-nitrophenyl)-2'-aminoethylsulfonate; PCMBS, para-(chloromercuri)benzene sulfonic acid.

Ian G. Macara and Lewis C. Cantley ● Department of Biochemistry and Molecular Biology, Harvard University, Cambridge, Massachusetts 02138.

1976; Gunn, 1978; Cabantchik *et al.*, 1978; Steck, 1978; Knauf, 1979; Wieth and Brahm, 1983), the present chapter will concentrate on the most recent developments in the field and avoid historical discussion. In particular, we will assume, without detailed citation of the evidence, that exchange of inorganic anions across the red cell membrane is mediated exclusively by the 95-Kd Band 3 protein, of which there are about 1.2×10^6 copies per cell; that the mechanism of exchange can best be described kinetically by a "ping-pong" rather than a "simultaneous" model; and that exchange is very tightly coupled and electrically silent.

2. PURIFICATION AND RECONSTITUTION

Hemoglobin-free red cell membranes are easily prepared by osmotic lysis and repeated washing with 5 mM sodium phosphate, pH 8.0 (Kant and Steck, 1974). Such membranes contain both integral and peripheral proteins. The latter can be removed by washing the membranes in a medium of high ionic strength, which selectively elutes Band 6 (Yu and Steck, 1975a), by addition of EDTA to elute spectrin (Bands 1 and 2) (Fairbanks *et al.*, 1971), or by exposure to extremes of pH (Grinstein *et al.*, 1979), which effectively "strips" the membrane of all peripheral proteins. The last treatment provides Band 3-enriched membranes already about 70% pure of contaminating proteins. Band 3 may be selectively extracted from the membranes with 0.5% Triton X-100 (Yu and Steck, 1975a; Wolosin, 1980; Cabantchik *et al.*, 1980; Köhne *et al.*, 1981; Lukacovic *et al.*, 1981). Contaminating glycophorin can be removed by preextraction with a lower concentration (0.05%) of detergent (Köhne *et al.*, 1981).

Further purification can be accomplished by anion exchange chromatography (Yu and Steck, 1975a; Lukacovic *et al.*, 1981) followed by affinity chromatography using an activated thiol gel (Fukuda *et al.*, 1978) or a (*p*-(chloromercuri)-benzamido)ethyl agarose gel (Lukacovic *et al.*, 1981). The final preparations are at least 95% pure Band 3. Protein aggregation on storage can be prevented by addition of 15 mM mercaptoethanol (Lukacovic *et al.*, 1981).

Purified Band 3 has been reconstituted successfully into phospholipid vesicles by a number of groups (Ross and McConnell, 1978; Cabantchik *et al.*, 1980; Wolosin, 1980; Köhne *et al.*, 1981; Lukacovic *et al.*, 1981). The best reconstitutions have used Band 3 that is at least 95% pure (Lukacovic *et al.*, 1981). Anion efflux from the vesicles was shown to be sensitive to inhibition by stilbene disulfonates and partially to require the presence of external, exchangeable anions. Several qualitative differences between the reconstituted and *in vivo* systems remain, however. In particular, reconstituted Band 3 displays a considerably reduced affinity for stilbene disulfonates and for sulfate (Cabantchik *et al.*, 1980; Köhne *et al.*, 1981), suggesting that the protein is no longer precisely in its native state, either because of the unphysiological phospholipid milieu of the vesicles or because of partial denaturation during purification. Even after the initial preparation of red cell ghosts, two sulfhydryl groups on Band 3 show reduced reactivity to *N*-ethyl maleimide (Rao, 1979; Ramjeesingh *et al.*, 1982), and changes in the behavior of the protein toward inhibitors occur upon storage of ghosts at 4°C (Macara and Cantley, 1981b), so that partial denaturation appears to be a real possibility.

An interesting alternative to reconstitution into phospholipid vesicles is the insertion of Band 3 into the plasma membranes of intact cells. Such an insertion has been achieved successfully by Volsky *et al.* (1979) using Band 3–Sendai virus envelopes and Friend erythroleukemia cells. Friend cells were chosen since they display a very low permeability to anions such as Cl^- (Harper and Knauf, 1979) and retain viability after fusion with the viral envelopes. The modified cells showed a large increase in stilbene-disulfonate-sensitive anion exchange. The technique should prove useful not only for the further investigation of isolated transport systems but also in cell biology as a means of introducing highly specific modifications to cell permeabilities.

3. STRUCTURE OF BAND 3

On SDS–polyacrylamide gel electrophoresis of erythrocyte membranes, Band 3 appears as a broad, diffuse band of 90–100 Kd (Fairbanks *et al.*, 1971). Purification does not improve the sharpness of the band (Yu and Steck, 1975a), and the apparent heterogeneity has been ascribed to a variable carbohydrate content (Jenkins and Tanner, 1977; Drickamer, 1978; Markowitz and Marchesi, 1981). The carbohydrate is present only on the C-terminal, membrane-bound portion of the protein that can be separated from a water-soluble, 40-Kd cytoplasmic portion by limited proteolysis (Jenkins and Tanner, 1977; Steck *et al.*, 1976) (Figure 1). These two sections of the protein

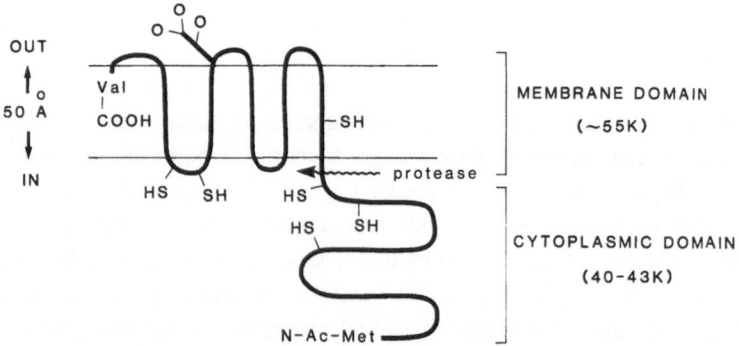

FIGURE 1. Organization of the Band 3 polypeptide in the red cell membrane, showing its two-domain structure.

appear to form distinct structurally independent domains with entirely separate and unrelated functions (Appell and Low, 1982). They will therefore be discussed separately.

3.1. The Cytoplasmic Domain

The N-terminal cytoplasmic domain can be purified after limited proteolysis by low concentrations of chymotrypsin that cleave Band 3 at only one location. Cyanylation or treatment with trypsin cuts the domain further, generating an N-terminal fragment of 22–23 Kd and one of 16–20 Kd (Steck *et al.*, 1976; Fukuda *et al.*, 1978) (Figure 2). The 23-Kd fragment has a blocked methionine residue (Drickamer, 1978), and of the first 31 residues on the N-terminal sequence, 16 are acidic and none are basic (Drickamer, 1978; Murthy *et al.*, 1981a). This negatively charged segment appears to provide a binding site for aldolase and glyceraldehyde-3-phosphate dehydrogenase (Murthy *et al.*, 1981b). The remainder of the 23-Kd peptide is relatively hydrophobic (Drickamer, 1977; Steck *et al.*, 1978).

The 20-Kd fragment of the cytoplasmic domain contains three of the five reactive sulfhydryl groups present in Band 3, the other two being in the membrane-bound domain (Rao and Reithmeier, 1979). Two of the three SH groups are very close to the two ends of the fragment, and the third is about 5 Kd from the trypsin-sensitive connection with the membrane domain (Figure 2). The 20-Kd fragment can be cleaved with cyanogen bromide into three

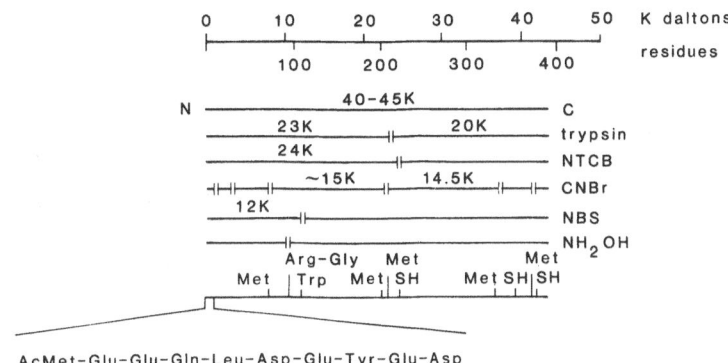

AcMet–Glu–Glu–Gln–Leu–Asp–Glu–Tyr–Glu–Asp

FIGURE 2. Primary structure of the cytoplasmic domain of Band 3. The data summarized in this figure were obtained from several sources, which are cited in the text. NTCB, 2-nitro-5-thiocyanobenzoic acid; CNBr, cyanogen bromide; NBS, *N*-bromosuccinimide; NH₂OH, hydroxylamine.

major peptides, each of which contains one of the three SH groups Rao, 1979; Rao and Reithmeier, 1979). Treatment of red cell membranes or Triton X-100-solubilized Band 3 with Cu^{2+}-*o*-phenanthroline results in cross-linking of the protein through the cytoplasmic domain sulfhydryls, to form covalent dimers. Surprisingly, only one disulfide bond is formed per dimer, but all three sulfhydryls can participate in cross-linking, a result that suggests first that the formation of one disulfide bond from any pair prevents the formation of others, perhaps because of a conformational change, and second that all six sulfhydryl groups on the cytoplasmic domains of the native Band 3 dimer must be near the dimer interface (Reithmeier and Rao, 1979).

Sedimentation analysis of the isolated cytoplasmic domain suggests a highly elongated morphology (Appell and Low, 1981), in agreement with electron micrographic evidence (Weinstein *et al.*, 1978), and the circular dichroic spectrum of the fragment corresponds to an α-helical content of about 37% (Appell and Low, 1981).

3.2. The Membrane-Bound Domain

The C-terminal, 55-Kd membrane-bound domain of Band 3 can be cleaved into two fragments (35 and 17 Kd; Figure 3) by incubation of intact cells with high concentrations of chymotrypsin (at 25–37°C) prior to lysis and

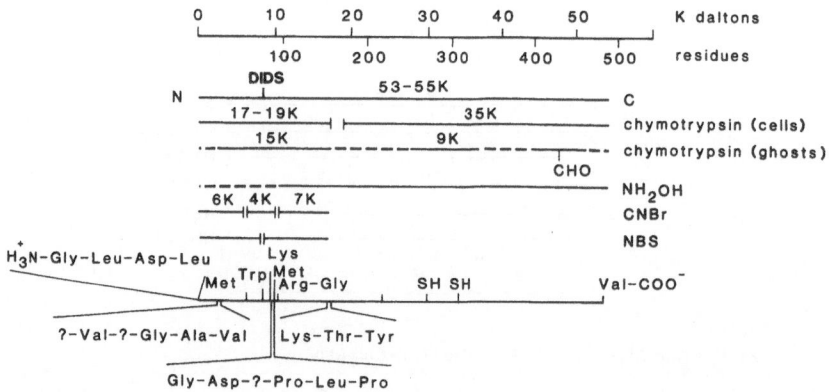

FIGURE 3. Primary structure of the membrane-bound domain. The data summarized in this figure were obtained from several sources, which are cited in the text. Abbreviations used are given in the caption to Figure 2.

removal of the cytoplasmic domain (Steck *et al.*, 1976; Drickamer, 1976; Grinstein *et al.*, 1978). Neither fragment is released from the membrane, and the two remain associated with one another even after solubilization in non-ionic detergents such as Triton X-100 (Reithmeier, 1979).

The C-terminal 35-Kd fragment contains all the carbohydrate on Band 3, attached through a single asparagine residue (Drickamer, 1978; Tanner, 1979). It also contains two of the five reactive SH groups present in Band 3 (Rao, 1979; Rao and Reithmeier, 1979). These SH groups are unusually sensitive to oxidation. They become unreactive to *N*-ethylmaleimide upon lysis of the cells and regain reactivity only after treatment with 3–5% β-mercaptoethanol (Rao, 1979). Since they are located on the cytoplasmic side of the membrane (Rao, 1979; Ramjeesingh *et al.*, 1982) while the carbohydrate moiety is exclusively extracellular, the 35-Kd fragment must traverse the membrane at least once. The same conclusion has been drawn from other labeling studies (Markowitz and Marchesi, 1981; Williams *et al.*, 1979).

Treatment of intact cells with papain after chymotrypsin cleavage removes a peptide of 5–10 Kd from the 35-Kd fragment, presumably from the C terminus (Jennings and Passow, 1979). If unsealed membranes are treated with high concentrations of chymotrypsin (1–2 mg/ml at 37°C for 1 hr), further cleavage reduces the 35-Kd fragment to a 9-Kd peptide (Ramjeesingh *et al.*, 1980a). Remarkably, this peptide cannot be generated by incubation of inside-out vesicles with chymotrypsin, suggesting that the cleavage occurs at the external surface of the membrane but is completely inhibited by the high ionic strength required to maintain the integrity of intact cells. This puzzling observation could be confirmed by comparing the susceptibilities of rightside-out and inside-out vesicles to the protease at different salt concentrations. The 9-Kd peptide does not contain the carbohydrate moiety present on the parent fragment, but it possesses the two reactive SH groups. Thus, if the proteolytic cleavages that generate the peptide are extracellular, the peptide must span the bilayer twice. Consistent with this topology is the observation that, although it is not hydrophobic, the peptide remains tightly associated with the membrane and is only slowly solubilized by nonionic detergents such as Triton X-100 (Ramjeesingh *et al.*, 1980a, 1982).

The 17-Kd fragment of the membrane-bound domain has received considerable attention because it can be labeled by affinity probes for the transport site of Band 3. Chymotrypsin (at 2 mg/ml) removes a 2-Kd, hydrophilic peptide from the N-terminal end of the fragment, presumably from the cytoplasmic side of the membrane (Ramjeesingh and Rothstein, 1983). Following purification by SDS electrophoresis and acetone precipitation, two further cleavages can be obtained by incubation with cyanogen bromide, as shown in Figure 2B (Ramjeesingh and Rothstein, 1983). A 4-Kd peptide from the middle of the fragment contains the lysine residue that reacts with affinity probes of the external transport site of Band 3.

Since the 15-Kd fragment is produced by cleavages at both sides of the membrane, it must span the bilayer at least once. (Assuming an α-helical arrangement, the minimum number of residues required to span a bilayer is about 28, creating a peptide of ~3 Kd.) Again, however, the amino acid composition of the fragment is not unusually hydrophobic (Ramjeesingh *et al.*, 1980b). One cryptic cysteine residue can be detected in the fragment, and it is probably contained in the 7-Kd peptide.

To complement the identification of exposed regions of Band 3, lipid-soluble reagents have been used to label specifically those parts of the protein that interact with the membrane bilayer (Wells and Findlay, 1980; Guidotti, 1980). The most informative results so far have been obtained by the use of a new reagent, [³H]adamantane diazirine, which has been developed by Knowles and co-workers (Bayley and Knowles, 1979). It can be photolysed at wavelengths well removed from those absorbed by proteins and produces a highly reactive, lipophilic carbene. As expected, most of the label appears on the 17-Kd and 35-Kd fragments of the membrane-bound domain of Band 3, with very little on the 40-Kd cytoplasmic domain. Surprisingly, however, the label on the 17-Kd fragment is located almost exclusively in the N-terminal half of the peptide. An 11-Kd peptide from the C-terminal end of the 17-Kd fragment, which can be labeled specifically with transport site affinity probes, does not react with the adamantane diazirine and presumably is not therefore in direct contact with membrane lipid fatty acid chains (Guidotti, 1980).

3.3. Transport Site Structure

3.3.1. Information from Limited Proteolytic Cleavages

The ability of Band 3 to exchange anions across the red cell membrane must reside exclusively in the membrane-bound domain of the protein, since removal of the cytoplasmic domain by proteolytic cleavage has no effect on transport (Grinstein *et al.*, 1978). Cleavage of the membrane-bound domain by extracellular chymotrypsin has no effect on transport either (Grinstein *et al.*, 1978), but the fragments remain tightly associated in the membrane (Reithmeier, 1979) and the tertiary organization of the domain is probably undisturbed. Further degradation of Band 3 with papain, however, strongly inhibits anion exchange (Ku *et al.*, 1979). The enzyme removes 5–10 Kd from the 35-Kd fragment (Jennings and Passow, 1979) and a small peptide from the C terminus of the 17-Kd fragment (Jennings and Adams, 1981). Analysis of the effects of papain on the transport properties of Band 3 (Jennings and Adams, 1981) has revealed that the proteolytic treatment specifically inhibits the outward translocation step of anion exchange. Inward translocation is actually accelerated two- to threefold, and there appears to be no significant effect upon the affinity of the external transport site for substrates. The affinity for stilbene disulfonates, which are potent inhibitors of anion exchange, is apparently reduced about 12-fold, however. These results suggest that papain treatment interferes with the conformational change accompanying anion exchange but does not damage the substrate binding site on Band 3. They indicate also that some of the residues involved in the binding of stilbene disulfonates may play an important role in the translocation step. Not surprisingly, treatment of red cell ghosts with a high concentration of chymotrypsin, which causes much more extensive degradation of the membrane domain (to a 9-

Kd and a 15-Kd fragment), also blocks anion transport (DuPre and Rothstein, 1981).

3.3.2. Information from Specific Inhibitors of Anion Exchange

The stilbene disulfonates and other aromatic sulfonates have proved very useful as affinity probes of Band 3 (for reviews, see Cabantchik et al., 1978; Knauf, 1979), because they can be used as reversible competitive inhibitors of anion exchange as well as covalent labels. They inhibit exchange exclusively from the external side of the membrane and react very specifically, with a stoichiometry of one molecule per Band 3 subunit. It is likely that they react with a residue located in the immediate neighborhood of the transport site, and stilbene disulfonates can therefore provide useful information about the position and environment of the site.

Proteolytic cleavage of Band 3, after labeling of cells with tritiated 4,4′-diisothiocyano-2,2′-dihydrostilbene disulfonate (^{3}H$_2$DIDS), identified the 17-Kd transmembrane fragment as the primary target of the affinity probe (Grinstein et al., 1978). More recent studies have localized the site of labeling to a lysine residue near the center of the fragment (Ramjeesingh et al., 1980b; Ramjeesingh and Rothstein, 1983).

Although H$_2$DIDS possesses two identical isothiocyano groups, only one normally reacts with Band 3. However, if the pH of the medium is raised to 9.5, the other also reacts and can cross-link the 17-Kd and 35-Kd fragments of the membrane-bound domain produced by chymotrypsin cleavage (Jennings and Passow, 1979). Pretreatment with papain before increasing the pH prevents the cross-linking. These results confirm the close association between the two fragments (the length of the H$_2$DIDS molecule is ~20 Å). They also demonstrate that every subunit of Band 3 can react with at least one molecule of H$_2$DIDS and indicate that the 35-Kd fragment as well as the 17-Kd fragment may be involved in anion exchange by Band 3.

Highly fluorescent derivatives of the stilbene disulfonates can be synthesized (Kotaki et al., 1971; Rao et al., 1979), and these have been used recently both to examine the kinetics of inhibitor binding (Dix et al., 1979) and to determine the position of the inhibitor binding site by fluorescence resonance energy transfer (Rao et al., 1979). Although 4,4′-diisothiocyanostilbene-2,2′-disulfonate (DIDS) is fluorescent, it has a low quantum yield (~0.03) and is unstable when bound to Band 3 because of the second isothiocyano group, which is available to react with water or with nearby lysine residues on the protein. 4-Benzamido-4'-isothiocyanostilbene-2,2'-disulfonate (BIDS) is subject to neither of these problems and is an excellent energy transfer donor, with a quantum yield when bound of 0.16 (Rao et al., 1979).

Resonance energy transfer was used to measure the distance between BIDS and fluorescent maleimides that had been reacted with the three sulfhydryl groups on the cytoplasmic domain of Band 3. Efficiencies of transfer were measured by sensitized emission, donor quenching, and donor lifetime changes (for background review, see Stryer, 1978). The distance between the stilbene disulfonate and the maleimides was determined to be 34–42 Å. Since this distance is less than the width of the bilayer, the anion binding site is probably located in a protein cleft that extends some way into the membrane. Polarization measurements demonstrated that the bound stilbene disulfonate is unable to rotate freely. The large fluorescence enhancement that occurs on binding is indicative of a hydrophobic environment, and quenching of the intrinsic protein fluorescence by stilbene disulfonates suggests that there are tryptophan residues near the inhibitor binding site (Rao et al., 1979). [There are probably four tryptophans present in the membrane-bound domain of Band 3 (Steck et al., 1976).] A hydrophobic environment is consistent with recent structure–activity studies, which have demonstrated a strong correlation between inhibitory potency and the hydrophobicity of the inhibitor (Barzilay et al., 1979; Sigrist et al., 1980).

Fluorescence studies using 4,4'-dibenzamidostilbene-2,2'-disulfonate (DBDS) and a novel stilbene disulfonate derivative, 4,4'-di(4-nitrobenz-2-oxo-1,3-diazole)-2,2'-dihydrostilbene disulfonate [$H_2(NBD)_2DS$], have revealed a negative cooperativity in binding to the external transport site of Band 3 (Dix et al., 1979; Macara and Cantley, 1981a). The affinity of $H_2(NBD)_2DS$ is tenfold lower to dimers containing a subunit already occupied by BIDS than to "empty" dimers. The BIDS has no effect, however, on the K_m for sulfate exchange, indicating that the negative interactions are the result of steric hindrance between the inhibitor molecules rather than allosteric effects. It has been suggested, therefore, that Band 3 forms an aqueous pore between its subunits that allows access to the transport sites (Macara and Cantley, 1981a).

A second type of anion exchange inhibitor that has proved useful in studying the anion binding sites of Band 3 is N-(4-azido-2-nitrophenyl)-2'-aminoethylsulfonate (NAP-taurine). In the dark, NAP-taurine can act both as a substrate for anion exchange and as a reversible inhibitor. On photolysis it converts to a reactive nitrene that covalently labels Band 3 with relatively high specificity (Cabantchik *et al.*, 1976). Unlike the stilbene disulfonates, NAP-taurine can inhibit anion exchange from either side of the membrane, competitively from the cytoplasmic side (K_I = 730 μM) and noncompetitively (K_I = 20 μM) from the extracellular side (Knauf *et al.*, 1978a). There is approximately one transport-related, high-affinity (10–20 μM) NAP-taurine binding site per Band 3 subunit (Macara and Cantley, 1981b). Kinetic data (Knauf *et al.*, 1978a) suggest that external NAP-taurine binds not to the anion substrate site but rather to a so-called "modifier site" that has been invoked in order to explain self-inhibition of equilibrium change at high anion concentrations (Dalmark, 1976; see Section 4.2.1b). Nonetheless, external [^{35}S]-NAP-taurine covalently labels the same 17-Kd transmembrane peptide as does ^{3}H$_2$DIDS (Knauf *et al.*, 1978b), and pretreatment with DIDS substantially reduces the labeling of Band 3 by either external or internal [^{35}S]-NAP-taurine (Cabantchik *et al.*, 1976; Grinstein *et al.*, 1979). Both equilibrium binding and kinetic studies recently have shown the reversible binding of NAP-taurine and stilbene disulfonates to be mutually exclusive (Macara and Cantley, 1981b; Fröhlich and Gunn, 1982), suggesting that these probes may bind to overlapping sites.

3.3.3. Information from Chemical Modification of Band 3

Traditional amino acid modification reagents have not as yet provided much information about active site residues in Band 3, possibly because very few of the residues are readily accessible, or because there are not many reactive residues near the anion transport site.

Certain reagents are "nonspecific" in that they label most of the proteins of the red cell membrane including Band 3, but they produce, nonetheless, specific and interesting effects on anion exchange that can be utilized to study the mechanism of transport. Thus, dansylation of resealed red cell ghosts with dansyl chloride has been shown to enhance SO_4^{2-} exchange by Band 3 (Legrum *et al.*, 1980), apparently by converting the "divalent" H^+/A^{2-} cotransport form of the carrier to the "monovalent" form. The dansyl chloride does not appear to react at the transport site of the protein, however, and does not block ^{3}H$_2$DIDS binding. Dinitrophenylation of red cells is also relatively "nonspecific," but, unlike treatment with dansyl chloride, it irreversibly inhibits anion exchange and completely blocks the reaction of ^{3}H$_2$DIDS with Band 3

(Zaki et al., 1975; Passow et al., 1980a), possibly by the modification of the same lysine residue (in the 17-Kd membrane-bound fragment) through which $^3H_2DIDS$ is attached. However, no equilibrium binding of dinitrofluoroben-zene (DNFB) to Band 3 is detectable prior to covalent reaction with the protein, and Passow and co-workers have used this property of the reagent, together with its ability to block reaction of $^3H_2DIDS$ with Band 3, to monitor changes in the accessibility of the DNFB-reactive lysine residue in response to a large number of different inhibitors of anion exchange (see Section 4.2.1d) (Passow et al., 1980a,b).

Although DNFB and stilbene disulfonates can react with lysine residues near the transport site, there has until very recently been no good evidence for the direct involvement of lysines in anion exchange, and titrations of the transport activity of resealed ghosts have shown only small effects on Cl⁻ exchange over the range within which deprotonation of a lysine is to be expected (Funder and Wieth, 1976). Transport is inhibited at very alkaline pHs, however, with a $pK_{1/2}$ of 12 at physiological ionic strength (Wieth and Bjerrum, 1982), suggesting involvement of an arginine residue in transport. This conclusion is supported by the observation that the arginine-specific reagent, phenylglyoxal, can irreversibly inhibit anion exchange (Wieth et al., 1982). At 0°C and physiological pH, inhibition is reversible but is limited to 50% of control values. Above pH 8.5, the extent of maximal inhibition increases toward 100%. Irreversible inhibition is achieved by raising the temperature (38°C, pH 10), and nonspecific labeling of membrane proteins is reduced by maintaining the intracellular pH near neutrality. Reaction of phenylglyoxal with the transport-related residues on Band 3 occurs very quickly (within a few seconds) under these conditions. High extracellular chloride or the presence of 4,4'-dinitro stilbene-2,2'-disulfonate (DNDS) both reduce the rate of reaction. Certain other potent inhibitors of anion exchange, however, including dipyrimadole and niflumic acid, have no effect upon the rate. Since these inhibitors, unlike DNDS, are noncompetitive with substrate anions, the results imply that the phenylglyoxal is reacting with essential arginine residues at or very near the external transport site of Band 3. The possibility is strength-ened by the observation that the pK_a of the group(s) with which the reagent reacts is about the same as that determined by direct titration of the anion transport rate. The effect of phenylglyoxal modification on equilibrium bind-ing of stilbene disulfonates has not yet been determined, but the reagent was shown to reduce the rate of reaction with DIDS fourfold, and at low DIDS concentrations (equimolar with Band 3 subunits), the maximum labeling of transport sites was reduced to 50% (Wieth *et al.*, 1982; Bjerrum *et al.*, 1982a). Zaki (1981), in a preliminary study on the inhibition of anion transport by another arginine-specific reagent (1,2-cyclohexanedione), reports that complete inactivation of transport by the reagent also reduced H_2DIDS binding by only about 50%, suggesting some form of interaction between the subunits of the Band 3 dimer. Interestingly, although DIDS labels the 17-Kd transmembrane fragment, phenylglyoxal reacts specifically with residues on the 35-Kd fragment, and maximum inactivation of transport is obtained by the binding of sufficient phenylglyoxal to modify only a single arginine residue in Band 3 (Bjerrum *et al.*, 1982a,b). These results strongly support earlier, circumstantial evidence for the direct involvement of the 35-Kd fragment in anion exchange.

Further information about residues at the transport site has been supplied recently by the modification of specific lysine residues on Band 3 by reductive methylation (Jennings, 1982a,b). This type of modification is unusually benign since it can be performed under mild conditions, does not alter charge, and introduces only minor steric perturbations. However, methylation of lysines prevents reaction with isothiocyanates, and Jennings has demonstrated that the reductive methylation of certain Band 3 lysine residues completely prevents the covalent labeling of Band 3 by H_2DIDS. This treatment also inhibits equilibrium Cl^- exchange by 75% but has no effect upon reversible H_2DIDS or DNDS binding and does not appear to alter the apparent affinity for extracellular substrate anions. Under the conditions used (pH 8.5, 0°C), only about two lysine residues per Band 3 subunit appear to be methylated; half of the label is on a single lysine residue in the 35-Kd transmembrane fragment and the remainder appears in the 17-Kd fragment. Presumably the 35-Kd lysine is the same residue that reacts with H_2DIDS to cross-link the fragments, although this has not yet been conclusively proven by sequence analysis. The methylation of the 17-Kd fragment is less specific. Only about 20% appears to be accounted for by methylation of the residue that reacts with H_2DIDS. Another 30–40% is very near the C terminus of the 17-Kd fragment and can be removed by papain treatment (Jennings, 1982a). Inhibition of Cl^- exchange correlates well with the methylation of the single lysine in the 35-Kd fragment. Another, hydrophobic inhibitor of anion exchange, phenylisothiocyanate, also reacts with a residue (probably lysine) on the 35-Kd peptide (Kempf *et al.*, 1981). Both the 17-Kd and 35-Kd fragments of the membrane-bound domain therefore appear to be intimately involved in the anion exchange mechanism.

3.3.4. Information from pH Titrations of Anion Exchange Activity

Early studies of the effect of pH demonstrated that anion exchange required the deprotonation of a group or groups on the red cell membrane with a pK_a of about 6.0. Above pH 8, there is little effect on Cl^-–Cl^- exchange at 0°C until pH 11 (Funder and Wieth, 1976), although at 37°C a small decrease in rate occurs (Brahm, 1977). However, sulfate exchange is rapidly inhibited above pH 6.5 (Schnell et al., 1977). These results have been explained in terms of titratable carrier model by Gunn (for review, see Gunn, 1978). Support for the model was supplied by the detection of sulfate–proton cotransport in Cl^-–SO_4^{2-} exchange experiments (Jennings, 1976). More recent studies have indicated that, although protons are obligatory for sulfate transport, the anion can bind to the unprotonated, outward-facing carrier and that the order of H^+ and SO_4^{2-} binding is random (Milanick and Gunn, 1982a). In the absence of sulfate, the pK_a of the protonatable group is 5.0, and the affinity for protons is increased about tenfold by bound sulfate. Conversely, protonation increases the apparent affinity for sulfate tenfold. Further information on the mechanism is likely to emerge from studies on dansylation of Band 3, which significantly enhances SO_4^{2-} exchange, apparently by "uncoupling" H^+/A^{2-} cotransport (Legrum et al., 1980). Both internal and external protons are able to inhibit Cl^- exchange (Wieth et al., 1980; Milanick and Gunn, 1982a,b), external protonation probably being at the same site as is required for sulfate influx. The pK_a of the internal group is 6.1 (Wieth et al., 1980).

Above pH 10, other classes of groups critical for Cl^- exchange are titrated (Wieth and Bjerrum, 1982), one with a pK_a of ~11, which appears to be related to the modifier site, and another with a pK_a of ~12 (at chloride concentrations >0.1 M), the protonation of which is absolutely required for transport activity. Both are likely to be arginine residues. The deprotonation of the first group(s) can stimulate transport rates under conditions in which the modifier site normally would be occupied (in the presence of high extracellular chloride or NAP-taurine concentrations, for instance). This observation suggests that the site can only be occupied by an inhibitory anion when protonated. It is not known if protonation of the transport-related group is required for substrate binding or only for translocation. The pK_a of the group is sensitive to external chloride, with a $K_{1/2}$ of about 16.5 mM. It is not clear, however, whether this effect is the result of the binding of Cl^- at the transport site or of a more general electrostatic interaction. Competition experiments in which iodide replaced the chloride were performed by Wieth and Bjerrum (1982) to distinguish these possibilities. Their results, however, are inconclusive in the light of recent evidence that the apparent differences in affinity of the Band 3 transport site for halides are a consequence largely of differences in translocation rates and that the true affinities for substrate anions are all rather similar (Milanick and Gunn, 1981; Knauf and Mann, 1982). Moreover, it is not certain that the titratable group is directly involved at the active site of Band 3. Deprotonation at unrelated sites could induce conformational changes that would be transmitted to the active site region. Modification studies strongly support the involvement of an arginine residue in anion exchange, however (Wieth *et al.*, 1982; Bjerrum *et al.*, 1982a,b; Zaki, 1981), and the effects of chloride concentration and pH suggest that the essential arginine being modified is the same group as was revealed by titration of transport activity.

3.4. Quaternary Structure of Band 3

It has been known for some years that Band 3 behaves in detergent solution and in the red cell membrane as a stable, noncovalent dimer (Yu and Steck, 1975b; Steck *et al.*, 1976; Reithmeier, 1979; Nigg and Cherry, 1979). Both domains appear to be involved in dimer formation. The isolated membrane-bound domain remains tightly associated even on solubilization in Triton X-100 (Reithmeier, 1979). The isolated cytoplasmic domain also behaves as a dimer in solution (Appell and Low, 1981), although it can no longer be cross-linked by Cu^{2+}-*o*-phenanthroline (Reithmeier, 1979). Recently, the distance between the stilbene disulfonate sites on adjacent subunits of Band 3 has been determined by resonance energy transfer measurements to be only 28–45 Å, which is considerably less than the calculated mean distance between unassociated subunits in the red cell membrane (~75 Å) (Macara and Cantley, 1981a). The distance was unaffected by solubilization in Triton X-100, indicating that the detergent-solubilized protein has a transport site structure similar to that observed in the membrane.

Surprisingly, dimers alone could not account fully for the resonance energy transfer data, and it was suggested that tetramers (or higher oligomers)

of Band 3 occur in the membrane (Macara and Cantley, 1981a). Both dimers
and tetramers have been observed by electrophoresis in nonionic detergents
(Nakashima et al., 1981), and the examination of intramembrane particles of
red cell membranes by electron microscopy has also suggested the existence
of tetramers (Weinstein et al., 1980). The particles were identified as Band
3 by selective proteolysis (Weinstein et al., 1978) and by their ability to bind
glyceraldehyde-3-phosphate dehydrogenase (GPDH), but the total population
was estimated to be only $3.6–4.5 \times 10^5$/cell, whereas the Band 3 subunit
population is very close to 1.2×10^6/cell (Knauf, 1979). Further evidence
for tetramers has been supplied by sedimentation analysis of Band 3 preparations solubilized in Ammonyx-LO (Pappert and Schubert, 1982) or in 92%
acetic acid and diluted into low-ionic-strength acidic or alkaline solutions
(Dorst and Schubert, 1979). These preparations also appear to contain Band
3 monomers at high dilution (<200 μg/ml), but the drastic conditions used
for solubilization make the results of dubious physiological significance. We
have been unable to prepare hybrid dimers of BIDS-labeled and eosin-mal-
eimide-labeled Band 3 by solubilization of mixtures of individually labeled
membranes in Triton X-100 (J. Lytton and I. G. Macara, unpublished observation) even after incubations at 25°C for 24 hr. Monomer–dimer equilibria
are therefore negligibly slow, at least in this detergent. Nor are they likely
to be significant in vivo, since the Band 3 polypeptide concentration (treating
the red cell membrane as a discrete volume 100 Å deep and of 100 μm²
surface area) is extremely high: about 3.5 mM or 340 mg/ml.

The overall shape of the Band 3 molecule is unknown. Electron micro-
graphs of freeze-fractured red cell membranes have suggested that the protein
extends 40–50 Å out from the membrane on the extracellular surface and that
the intramembrane particles (tetramers?) are 66 Å in diameter at the fracture
face (Weinstein et al., 1978, 1980). Calculating from the molecular weight
of the membrane domain of Band 3, the polypeptide could occupy a cylinder
30 Å in diameter and 100 Å in length. The particles might therefore comprise
associations of such cylinders. However, recent measurements of resonance
energy transfer from the protein tryptophan residues to n-anthroyloxy fatty
acids in the outer leaflet of the membrane bilayer indicate that the diameter
of Band 3 is greater near the extracellular surface than near the intracellular
surface, so that its shape cannot be approximated accurately by a simple
cylinder (A. Kleinfeld, personal communication). The measurements also

indicate two distinct tryptophan-containing regions, one extending about 10 Å into the cytoplasm and the other near the outer surface of the membrane (Kleinfeld *et al.*, 1982).

3.5. Posttranslational Modifications of Band 3

3.5.1. Glycosylation

Virtually all of the carbohydrate on Band 3 is located on the 35-Kd fragment of the membrane-bound domain (Fukuda *et al.*, 1978; Drickamer, 1978). It probably comprises a single oligosaccharide chain per Band 3 monomer of about 5.5 Kd, with an average of about 30 sugar residues, and is attached through an asparagine residue. It can be cleaved by hydrazine into a high- and a low-molecular-weight oligosaccharide. The large fragment consists of a variable number of the repeating unit:

$$\text{Gal } 1 \overset{\beta}{\to} 4 \text{ GlcNAc } 1 \overset{\beta}{\to} 3$$

in the peripheral region (Tsuji *et al.*, 1980), together with a core structure of the following structure:

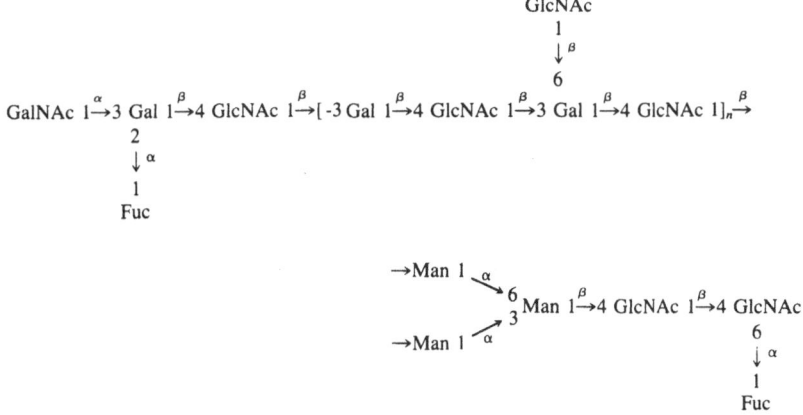

The smaller fragment also has a branched structure (Tsuji *et al.*, 1981):

$$\text{Gal } 1\overset{\beta}{\rightarrow}4 \text{ GlcNAc } 1\overset{\beta}{\rightarrow} \text{ Man} \qquad\qquad\qquad\qquad \text{Fuc}$$

$$2 \qquad\qquad\qquad\qquad\qquad\qquad\qquad\qquad 1$$

$$\downarrow \alpha \qquad\qquad\qquad\qquad\qquad\qquad\qquad \downarrow$$

$$6 \qquad\qquad\qquad\qquad\qquad\qquad\qquad\qquad 6$$

$$\text{GlcNAc } 1\overset{\beta}{\rightarrow}4 \text{ Man } 1\overset{\beta}{\rightarrow}4 \text{ GlcNAc } 1\overset{\beta}{\rightarrow}4 \text{ GlcNAc}$$

$$3$$

$$\downarrow \alpha$$

$$2$$

$$\text{Gal } 1\overset{\beta}{\rightarrow}4 \text{ GlcNAc } 1\overset{\beta}{\rightarrow} \text{ Man}$$

3.5.2. Phosphorylation

Phosphorylation of Band 3 by ATP appears to be confined to the cytoplasmic domain of the protein, and the major site for phosphorylation is in the polar 23-Kd fragment, within 100 residues of the N terminus (Drickamer, 1976, 1977). No phosphorylation was observed in the adjacent peptide, indicating that the labeling is specific. Minor phosphorylation sites occur in the 20-Kd C-terminal region of the cytoplasmic domain. The physiological significance of Band 3 phosphorylation is unknown. It would be of interest to know if it modulated the association with cytosolic proteins, since their binding sites are in the same region as the major sites of phosphorylation (Murthy *et al.*, 1981b).

3.5.3. Methylation

Recently it has been discovered that Band 3 can be methylated specifically both *in vivo* and *in vitro* under conditions that closely mimic those within the intact red cell (Terwilliger and Clark, 1981; Freitag and Clark, 1981). When packed erythrocytes were lysed by freeze–thawing in the presence of *S*-adenosyl[*methyl*-^{3}H]methionine, label was incorporated mainly into intrinsic membrane proteins, probably through reaction with aspartyl and glutamyl residues. In the intact cells, however, using L-[*methyl*-^{3}H]methionine, several cytoskeletal proteins (Bands 2.1 and 4.1) were also labeled. One site of methylation on Band 3 appears to be at the junction between the cytoplasmic and membrane-bound domains. Again, as for phosphorylation, the physiological significance of the modification, if any, is as yet unknown.

4. FUNCTIONS OF BAND 3

Band 3 performs several independent tasks in the human red cell, a fact reflected in its two-domain structure. The membrane-bound domain is concerned with transport; the cytoplasmic domain with protein–protein associations and the modulation of certain enzyme activities. It is not yet established whether or not the functions of the two domains interact.

4.1. Functions of the Cytoplasmic Domain

The cytoplasmic domain possesses two related functions that appear to be the concern of different "subdomains." One major, well-established task is to act as an anchor for the cytoskeleton. Another, of less certain physiological significance, is to bind glycolytic enzymes, and possibly to bind hemoglobin also. The binding site for the glycolytic enzymes is near the N terminus. Hemoglobin competes for this site, but the cytoskeletal proteins do not, and it is likely, but so far unproven, that the cytoskeleton is attached to the membrane via the C terminal 20-Kd fragment of the cytoplasmic domain of Band 3.

4.1.1. Associations with the Cytoskeleton

The red cell cytoskeleton forms a net, spread like the struts of an untidy geodesic dome over the entire surface of the membrane, to control the shape and elasticity of the cell. It can be prepared easily from red cell ghosts by extraction of the membrane lipids and intrinsic proteins with Triton X-100 (Yu *et al.*, 1973) and is composed mainly of the peripheral proteins spectrin (Bands 1 and 2), ankyrin (Band 2.1), Band 4.1, and actin (Band 5), together with some Band 3. Of these proteins only ankyrin binds specifically to the cytoplasmic domain of Band 3 (Hargreaves *et al.*, 1980; Bennett and Stenbuck, 1979, 1980). Spectrin, an elongated molecule that forms 75% by weight of the cytoskeleton, is attached to ankyrin. It possesses specific binding sites for Band 4.1 and actin and also self-associates into tetramers, thereby creating the closed net observed after detergent extraction of the membrane (for review see Branton *et al.*, 1981).

Purified ankyrin binds to "stripped" inside-out vesicles with an affinity of ~5 × 10^{-8} M under physiological conditions (Hargreaves *et al.*, 1980). Binding is very slow at room temperature ($t_{1/2} \sim$ 30 min), saturable, protease-sensitive, and competively inhibited by the addition of the purified cytoplasmic domain of Band 3. Curiously, the stoichiometry of high-affinity binding appears (after corrections for contamination, etc.) to be only 1.5–2.5 × 10^5 per cell or about one ankyrin per four to eight Band 3 monomers. There cannot be a subset of Band 3 molecules in the red cell that is unable to bind ankyrin, because purified cytoplasmic domain associates with a stoichiometry of 1 : 1 (Bennett and Stenbuck, 1980) and because those Band 3 molecules released from the cytoskeleton during Triton X-100 solubilization of ghosts are still competent to bind ankyrin (Hargreaves *et al.*, 1980). It is possible that if Band 3 occurs as tetramers *in vivo* then only a single ankyrin molecule per tetramer can be accommodated at the binding site. Since there are only 1–2 × 10^5 molecules of ankyrin per cell, less than one site per tetramer is required anyway (Branton *et al.*, 1981). From rotational diffusion measurements of Band 3 in erythrocyte ghosts, it has been calculated that only about 40% of Band 3 polypeptides may be linked to the cytoskeleton (Nigg and Cherry, 1980). This value is equivalent to about 1 × 10^5 tetramers per cell and is roughly equal to the number of ankyrin molecules. Lateral mobility measurements of Band 3 using the fluorescence photobleaching recovery technique suggest that the proportion of protein immobilized by interactions with the cytoskeleton under physiological conditions is closer to 60% (Golan and Veatch, 1980). Addition of ankyrin fragments to compete for binding to Band 3 or incubation of ghosts in high KCl concentrations to destabilize ankyrin–Band 3 associations both led to large (90%) increases in the fractional mobility, indicating that ankyrin is primarily responsible for restricting the lateral mobility of Band 3 in the red cell membrane (Golan and Veatch, 1982).

The large excess in copies of Band 3 over the number required for cytoskeletal interactions probably exists because the number of copies is determined primarily by the requirement for rapid Cl^-/HCO_3^- exchange via the transmembrane domain.

4.1.2. Association with Glycolytic Enzymes

Hemoglobin, together with certain glycolytic enzymes, has been known for many years to remain associated with red cell ghosts during preparation. Yu and Steck (1975b) showed that GPDH in particular (Band 6) copurified with Band 3 after extraction from membranes. The stoichiometry of binding is 1 : 1 (GPDH tetramer : Band 3 protomer), and the association displays positive cooperativity. The binding site is in the highly acidic N-terminal 23-Kd peptide (Table I) and is very probably the same site as binds aldolase (Tsai *et al.*, 1982). Ankyrin binding is unaffected by the presence of GPDH (Hargreaves *et al.*, 1980). The interaction with Band 3 is primarily electrostatic, but, although the enzyme can be eluted efficiently from membranes by high NaCl concentrations, alkaline pH, polyvalent anions, and NADH and 2,3-diphosphoglycerate, as much as 50–70% of the total GPDH in the red cell has been estimated to be bound to Band 3 *in vivo* (Kliman and Steck, 1980). This estimate was obtained in two ways: (1) by extrapolation from equilibrium binding measurements at physiological pH and ionic strength in the presence of varying amounts of red cell contents; and (2) by rapid filtration of detergent-solubilized cells and measurement of the time-dependent release of GPDH into solution. Extrapolation of release to the zero time intercept provides the amount of GPDH bound to the membrane at the instant of cell lysis, which is assumed to represent the amount bound *in vivo*.

Electron spin resonance studies have suggested that a conformational change occurs near the active sites of GPDH on binding to red cell membranes (Beth *et al.*, 1981) and the enzyme is completely inhibited by association with Band 3 (Tsai *et al.*, 1982). Inhibition by membranes is competitive with respect to NAD^+ but is noncompetitive with glyceraldehyde-3-phosphate.

The interaction of aldolase with Band 3 is apparently very similar to that of GPDH, the two enzymes competing for a common site on the N-terminal portion of the cytoplasmic domain. Aldolase binds with a stoichiometry of one tetramer per Band 3 protomer, is eluted by high ionic strength and substrate, and is inhibited when bound (Strapazon and Steck, 1976, 1977; Murthy *et al.*, 1981a,b). Rapid filtration studies indicate that about half of the total red cell aldolase is bound to Band 3 *in vivo* (J. D. Jenkins and T. L. Steck, personal communication) (Table I). Cross-linking studies, using glutaraldehyde, also have detected the *in vivo* association of aldolase with the red cell membrane (Yeltman and Harris, 1980), but the technique cannot

TABLE I. Association of Red Cell Proteins with the Cytoplasmic Domain of Band 3

Protein	Copies/cell	High-affinity sites/cell	Percent bound *in vivo*[a]	Location of binding site	K_d (physiological conditions)
Ankyrin	1–2×10^5	1.5–2.5×10^5	~100%	22 Kd (C terminus)	~1×10^{-8} M
GPDH	1.4–2.4×10^5	~1.2×10^6	50–70%	23 Kd (N terminus)	~3×10^{-5} M
Aldolase	1.2×10^5	~1.2×10^6	40%	23 Kd (N terminus)	?
Phosphofructokinase	1.5–3×10^4	~4×10^5	10–20%	?	?
Hemoglobin	3×10^8	~1.4×10^6	~0%	23 Kd?	?

[a] Data are taken from sources cited in the text. The *in vivo* binding data were obtained by the rapid filtration technique of Kliman and Steck (1980). The values for aldolase and PFK were provided by J. D. Jenkins and T. L. Steck (personal communication).

provide any information on the amount bound because, in the presence of a cross-linker, the system is no longer at equilibrium.

A preliminary investigation (Higashi *et al.*, 1979) suggests that phosphofructokinase, the rate-limiting enzyme in glycolysis, might also associate specifically with Band 3. Rabbit muscle phosphofructokinase was reported to bind at low ionic strength to a single class of sites on human red cell membranes with a stoichiometry of $\sim 4 \times 10^5$/cell and to be eluted by either aldolase or GPDH. Phosphofructokinase is not inhibited by binding to ghosts, however, but rather loses sensitivity to the allosteric effectors ATP, 2,3-diphosphoglycerate, and citrate (Karadsheh and Uyeda, 1977). Rapid filtration analysis has indicated that 10–20% of the red cell phosphofructokinase is bound to the membrane *in vivo* (J. D. Jenkins and T. L. Steck, personal communication). Unfortunately, since no phosphofructokinase : Band 3 complex could be demonstrated by sedimentation analysis, and competition binding studies with the purified cytoplasmic domain of Band 3 were not performed (Higashi *et al.*, 1979), the evidence for a specific association remains inconclusive.

What might be the function of an association between Band 3 and glycolytic enzymes? Regulation of cellular glycolysis is unlikely, since the activities of the inhibited enzymes are not rate-limiting; nor could binding modulate anion transport activity, because transport is unaffected by either removal of peripheral proteins from inside-out vesicles or removal of the cytoplasmic domain by proteolysis. One intriguing possibility is that the bound enzymes create a separate "membrane pool" of ATP that is utilized for active transport (Hoffman and Proverbio, 1974) and by membrane-bound kinases or the spectrin-dependent ATPase (Baskin and Langdon, 1981). In support of this hypothesis, Mercer and Dunham (1982) have very recently demonstrated the stimulation of Na^+,K^+-ATPase-driven Na^+ transport into inside-out red cell vesicles by the substrates for GPDH and phosphoglycerate kinase, in the absence of added ATP. Elution of GPDH blocked the stimulation. Fossel and Solomon (1977, 1979) have suggested that close associations occur between the Na^+,K^+-ATPase, Band 3, and GPDH on the basis of small shifts in the ^{31}P nmr signal from enzyme-bound 2,3-diphosphoglycerate upon addition of ouabain or DIDS. The problems with the "membrane ATP pool" hypothesis are first that the number of Na^+,K^+-ATPase units per cell is very small (250–500) compared with the amount of Band 3 (1×10^6/cell) and bound GPDH ($\sim 1 \times 10^5$/cell), and second that GPDH is inactive when bound to Band 3. It is possible, however, that only a small subset of membrane-bound glycolytic enzymes, attached to sites other than Band 3, supply ATP to the Na^+,K^+-ATPase.

4.1.3. Association with Hemoglobins

In spite of numerous studies on the interaction of hemoglobins with the red cell membrane, none published to date has established definitively either that the cytoplasmic domain of Band 3 is the predominant high-affinity binding site on the membrane or that significant binding occurs under physiological conditions.

The hemoglobin concentration in the red cell is about 5mM tetramer, but most binding studies have been performed at very low hemoglobin concentrations and low pH and ionic strength, under which conditions the hemoglobin exists predominantly as $\alpha\beta$ dimers. About $1-1.4 \times 10^6$ high-affinity hemoglobin sites per cell have been detected (Shaklai *et al.*, 1977a,b; Salhany *et al.*, 1980). Limited proteolysis reduced the binding capacity of the membranes and hemoglobin could be eluted from them by raising the pH or ionic strength. Addition of GPDH specifically reduced binding at the high-affinity sites (Salhany *et al.*, 1980), but the location of these sites on Band 3 was not definitively proven. However, in recent experiments from this laboratory, using the purified 43-Kd cytoplasmic domain of Band 3, specific binding of hemoglobin dimers was detectable even at pH 6.8 and in 159 mM KCl (J. Belsky, unpublished observations). The peptide has been labeled with a fluorescent maleimide, and bound hemoglobin quenched the emission by resonance energy transfer. Hemoglobin was competed off the labeled peptide by unlabeled material and by aldolase and GPDH. Recently Sayare and Fikiet (1981) have shown that hemoglobin can be cross-linked with Band 3 via the β-globin chain through intermolecular disulfide bonds.

Attempts have been made to detect hemoglobin binding to red cell membranes *in vivo* or at near-physiological hemoglobin concentrations (Eisinger *et al.*, 1982), but the problems attending such measurements are virtually insuperable, because at *in vivo* concentrations (~ 5 mM tetramer) the mean edge-to-edge distance between hemoglobin molecules, or between hemoglobin and the membrane, is only ~ 30 Å, which is much less than the mean separation between Band 3 dimers in the membrane. Furthermore, the total amount of Band 3 is less than 0.4% of the hemoglobin present in the cell.

The purpose of an association between hemoglobin and Band 3 remains obscure. Kinetic measurements of CO binding have suggested that the high-O_2-affinity conformation of hemoglobin is stabilized by the association (Salhany and Shaklai, 1979), and that oxyhemoglobin binds more tightly than the deoxy form. Competition for binding with glycolytic enzymes might therefore vary with oxygen status, but there is no evidence as yet for such an effect. Several investigators have attempted to demonstrate differences in affinity for red cell membranes among mutant hemoglobins, particularly HbS, which can produce irreversible sickling of erythrocytes and decreases in cell flexibility (Shaklai *et al.*, 1981; Premachandra and Mentzer, 1980; Schneider *et al.*, 1980). Because the β chains of HbS have a greater positive charge than those of HbA, one can predict that the mutant hemoglobin should have the higher affinity for the acidic cytoplasmic domains of Band 3, and this is in fact observed (Schneider *et al.*, 1980). There is no evidence, however, that the difference in affinity has any bearing on the sickle cell phenotype, and in a careful study by Goldberg and co-workers (1981) no effect of red cell membranes on the rate of HbS polymerization was detectable.

4.2. Functions of the Membrane-Bound Domain

The primary function of the membrane domain is to exchange Cl^- and HCO_3^- across the red cell membrane. There is little substrate specificity, however, and many other anions, both organic and inorganic, can be exchanged by Band 3. Water and nonelectrolytes such as urea may also be transported, and a very slow net flux of anions occurs through the protein. The transmembrane domain also appears to associate with another transmembrane protein, glycophorin A (Cherry and Nigg, 1980), as judged by the reduction in rotational diffusion of Band 3 produced by the addition of antiglycophorin A antibodies. There are sufficient glycophorin A molecules present in the red cell membrane to associate with Band 3 dimers in a 1 : 1 stoichiometry, but the purpose of such an interaction is unknown.

4.2.1. Mechanism of Anion Exchange

4.2.1a. The Ping-Pong Kinetic Model. All the available evidence to date supports a "ping-pong" kinetic model for the exchange of anions by Band 3 (for recent reviews see Cabantchik *et al.*, 1978; Knauf, 1979; for classical kinetic tests of the model see Gunn and Fröhlich, 1979). The model predicts that the transport protein cannot bind internal and external anions simultaneously (i.e., it behaves kinetically as a single rotating carrier, although physically separate internal transport sites are not excluded); that sites can be "recruited" to one side or other of the membrane by the imposition of anion gradients; that unoccupied sites alter their state only very slowly; and that a conformational change in the protein occurs twice during each exchange cycle. Most of these predictions have already been verified, but the molecular details of the exchange mechanism remain obscure.

The inability of Band 3 to bind internal and external anions simultaneously was demonstrated in a nonkinetic manner by Rothstein and co-workers (Grinstein *et al.*, 1979) using the transport site affinity probes DIDS and NAP-taurine. They showed that the reactions of DIDS at the out-facing transport site blocked the subsequent binding of $[^{35}S]$-NAP-taurine to Band 3 on the opposite site of the membrane. Although extracellular NAP-taurine appears to associate primarily with the so-called "modifier site" and behaves as a noncompetitive inhibitor, internal NAP-taurine is a competitive inhibitor of anion exchange, presumably binding at the in-facing transport site (Knauf *et al.*, 1978a). The experiment did not, of course, exclude the possible existence of two interacting transport sites of the type shown in Figure 4. Moreover, the large size and hydrophobicity of the DIDS molecule raise the possibility that its association with the out-facing transport site induces conformational changes in Band 3 unrelated to the transport mechanism.

Recent "site-recruitment" studies have provided less ambiguous support for the ping-pong model. Knauf and co-workers (1980) have shown, for instance, that a high $[Cl]_{in}/[Cl]_{out}$ gradient raises the apparent affinity of Band 3 for external H_2DIDS. A gradient of this type increases the proportion of out-facing sites, according to the ping-pong model, as described by the equation

$$\frac{E_{out}}{E_{in}} = \frac{k_{-1} \cdot K_{out}[Cl]_{in}}{k_1 \cdot K_{in}[Cl]_{out}} \tag{1}$$

where E_{out}/E_{in} is the ratio of out-facing to in-facing sites, k_1 and k_{-1} are the rate constants for the conversion $E_{out}Cl \rightarrow E_{in}Cl$ and vice versa, respectively, and K_{out} and K_{in} are the dissociation constants of the out- and in-facing sites. If there is less chloride outside the membrane, the carrier spends more time in the unloaded, out-facing state. Less H_2DIDS, which only interacts with the out-facing state, is then required to inhibit exchange (Knauf, 1979; Knauf et al., 1980).

Similar manipulations of chloride gradients by Jennings (1980) also have been shown to affect the initial (nonequilibrium) fluxes of $^{35}SO_4^{2-}$ across the membrane in a manner congruent with that predicted by the ping-pong mechanism. Furthermore, the net efflux of $^{36}Cl^-$ into a chloride-free phosphate medium was found to be independent of the intracellular chloride concentration over the range 20–170 mM (Jennings, 1982c). Phosphate is translocated much more slowly than chloride by Band 3, so that even at chloride concentrations lower than K_{in} the rate of Cl–Pi exchange is limited (in the ping-pong model) by the rate of the phosphate translocation. In a similar type of experiment, cells containing a very low concentration of chloride were resuspended in a completely chloride-free medium, and from the initial net efflux of the anion it could be calculated that only one chloride ion is transported with each half-cycle of exchange by the Band 3 polypeptide (Jennings, 1982c). This elegant experiment has provided the first estimate of the true stoichiometry of anion translocation by Band 3.

Investigation of the equilibrium exchange of chloride and bromide by Gunn and Fröhlich (1979) suggested that Band 3 is intrinsically asymmetric, with different apparent affinities for internal and external chloride (apparent K_{in} = 60 mM; apparent K_{out} = 4 mM). However, the experiments could not distinguish whether the asymmetry was the result of a true difference in binding affinity between the in- and out-facing sites or of a difference in the translocation rates, k_1 and k_{-1} [see equation (1)]. Whatever the cause of the asymmetry, the experiments imply that, even in the absence of any anion gradient, the numbers of in- and out-facing transport sites will not be equal. In fact, one can predict that E_{in}/E_{out} = 15 when Cl_{in} = Cl_{out}. Qualitative agreement with this prediction has been obtained from measurements of the inhibitory potency of H_2DIDS (Knauf et al., 1980). From similar measurements of the potency of niflumic acid, which is a strictly noncompetitive inhibitor of anion exchange (Cousin and Motais, 1979) that binds preferentially to E_{out} or ECl_{out}, the asymmetry has been ascribed entirely to differences in the rate constants for inward and outward translocation of anions, rather than to differences in binding affinities (Knauf and Mann, 1982). This result has important consequences for studies of conformational changes in Band 3 because it invalidates the common assumption (Passow et al., 1980a,b; Wieth and Bjerrum, 1982) that changing the equilibrium chloride concentrations will shift the E_{in}/E_{out} ratio. Of course, in the complete absence of substrate, the E_{in}/E_{out} ratio will no longer depend on k_{-1}/k_1 but on the ratio of the rate constants for the conformational change of the unloaded carrier (Passow *et al.*, 1980a,b). However, these rate constants are likely to be extremely small (see Section 4.2.2), and it is not clear whether the transporter will attain equilibrium or will rather remain "trapped" at the E_{in}/E_{out} ratio existing in the presence of substrate anions. The latter possibility could be tested by using the dinitrophenylation technique of Passow and co-workers (1980a,b) to determine the relative exposure of the conformation-sensitive lysine in substrate-free medium after exposure of the transporter to different anion gradients (see Section 4.2.1d).

4.2.1b. Substrate Inhibition and Allosteric Interactions. The ping-pong mechanism for anion exchange is complicated by the occurrence of substrate inhibition at high anion concentrations, an effect that has been ascribed to the existence of an externally located "modifier" site (Dalmark, 1976). Kinetic evidence suggests that extracellular NAP-taurine, a noncompetitive inhibitor, interacts preferentially with this site (Knauf *et al.*, 1978a). It is evident, however, that a modifier site is only one of several possible explanations of the data. Negative cooperativity between Band 3 subunits provides one plausible alternative. Clearly, if a separate, inhibitory site exists, it must be very close to the transport site, because [^{35}S]-NAP-taurine labels the same 17-Kd transmembrane fragment of Band 3 as $^3H_2$DIDS (Knauf *et al.*, 1978b), and the reversible binding of NAP-taurine is mutually exclusive not only with that of competitive inhibitors of anion exchange such as 4-benzamido-4'-amino-stilbene-2,2'-disulfonate and DNDS (Macara and Cantley, 1981b; Fröhlich and Gunn, 1982) but also, surprisingly, with sulfate, which is a transportable anion like chloride (Fröhlich and Gunn, 1982). Moreover, the high-affinity binding of NAP-taurine resembles that of stilbene disulfonates in being limited to Band 3 in the out-facing state (Knauf *et al.*, 1980). Negative cooperativity between subunits is contraindicated by the fact that the fractional inhibition of anion exchange is closely proportional to the fraction of Band 3 molecules that have been irreversibly inhibited by reaction with DIDS or H_2DIDS Lepke *et al.*, 1976; Ship *et al.*, 1977). Salhany and Gaines (1981) have nonetheless presented evidence in favor of a negative allosteric mechanism for the heteroexchange of dithionite ($S_2O_4^{2-}$) and sulfate by Band 3. Interpretation of their results is complicated by the chemical equilibrium between $S_2O_4^{2-}$ and SO_2^-, both of which presumably can act as substrates, and by an unknown contribution from the net fluxes of these anions. The apparent negative cooperativity can probably be ascribed to a changeover from dithionite-limited to sulfate-limited exchange at high dithionite concentrations. We have demonstrated recently that, although aromatic disulfonate binding displays a substantial negative cooperativity, the effect is a result of steric hindrance rather than allostery (Dix *et al.*, 1979; Macara and Cantley, 1981a). No change in the apparent affinity for sulfate or NAP-taurine is created by covalent reaction of a stilbene disulfonate (BIDS) with 70% of the Band 3 transport sites on the red cell membrane (Macara and Cantley, 1981b).

4.2.1c. A Model of the Band 3 Transport Site.   We will now present a speculative model of the transport site to facilitate further discussion of the transport site structure and mechanism of action of Band 3. The model should be regarded as a useful fiction which, although unlikely to be correct in detail, is able nonetheless to explain many of the observations described previously, and which readily lends itself to a plausible hypothesis for the anion exchange mechanism. The model is shown in Figures 4 and 5.

The essence of the model is that a positively charged cavity in the Band 3 protein alternately becomes exposed to the inside and outside of the cell because of the movement of an "anionic gate." The movement of the gate is driven by charge repulsion by the anion being transported. Two anion binding sites per Band 3 monomer are envisaged, access to each being controlled by the anionic gate (i.e., an aspartate or glutamate residue). Anion binding at these sites is stabilized by positively charged residues, which may alternately form salt bonds with the anionic gate (Figure 4). The model proposes that an incoming anion preferentially associates with the positively charged residue involved in the salt bond and by charge repulsion forces the gate to swing to the other positively charged group. The anion may now exit the cavity to the opposite side of the membrane, and the gate remains in the same position until a second anion binds from the side of the membrane where the previous anion exited. The sequence is then reversed (Figure 4).

The model explains successfully much of the available kinetic, titration, and chemical modification data on anion exchange by Band 3. The mechanism is clearly of the ping-pong type and will be tightly coupled because of the low probability that the anionic gate will change conformation in the absence of bound anion. Moreover, the model explains the inhibition of anion exchange by high substrate concentrations without recourse to a separate "modifier" site, since occupation of the two anion sites simultaneously will reduce

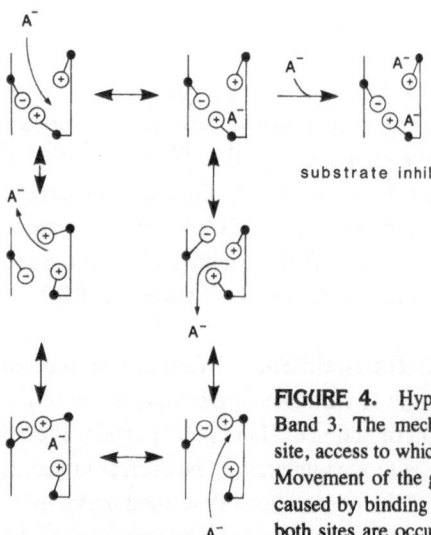

substrate inhibition

FIGURE 4. Hypothetical mechanism for anion exchange by Band 3. The mechanism involves an outer and an inner anion site, access to which is controlled by a negatively charged "gate." Movement of the gate is activated by the electrostatic repulsion caused by binding of an anion. Substrate inhibition occurs when both sites are occupied at the same time.

the transport efficiency by blocking the movement of the anionic gate (Figure 4). The model also predicts that molecules such as stilbene disulfonates or eosin maleimide, which possesses two anionic groups separated by ~12 Å, bind very tightly to Band 3 by occupying both anion sites simultaneously. It explains, furthermore, why NAP-taurine (which binds to the "modifier" site) and stilbene disulfonates (which probably span both sites) have different effects on anion exchange but show mutually exclusive binding (Figure 5).

The titration data described above suggest that the deprotonation of groups of pK_a 5–6, and the protonation of groups of pK_a 11–12, are necessary for anion transport by Band 3. Although there is no evidence for the direct participation of any of these groups in the active site of the protein, the titration data can be rationalized in terms of the model presented in Figures 4 and 5 if it is assumed that the positively charged groups that form salt bridges with the anionic gate are arginine residues and that the gate residue is an aspartate or glutamate. The pK_a of 5–6 is rather high for an acidic amino acid, particularly for one interacting with a nearby arginine, but it is not impossible in a hydrophobic environment. To explain the titration data of Wieth *et al.* (1980) and of Milanick and Gunn (1982a,b), the pK_a of the gate residue is taken to be about 5.0 when outward-facing and 6.1 when inward-facing (Figure 6A). Anion binding to the substrate site would partially screen the charged arginine residue nearby and raise the apparent pK_a. Protonation would inhibit movement of the gate because the electrostatic interaction with the arginines at the transport sites would be abolished. If the inhibition were not total, sulfate–proton cotransport might still occur, but Cl^- exchange, which is normally 10^4 times faster than sulfate exchange, would appear to be com-

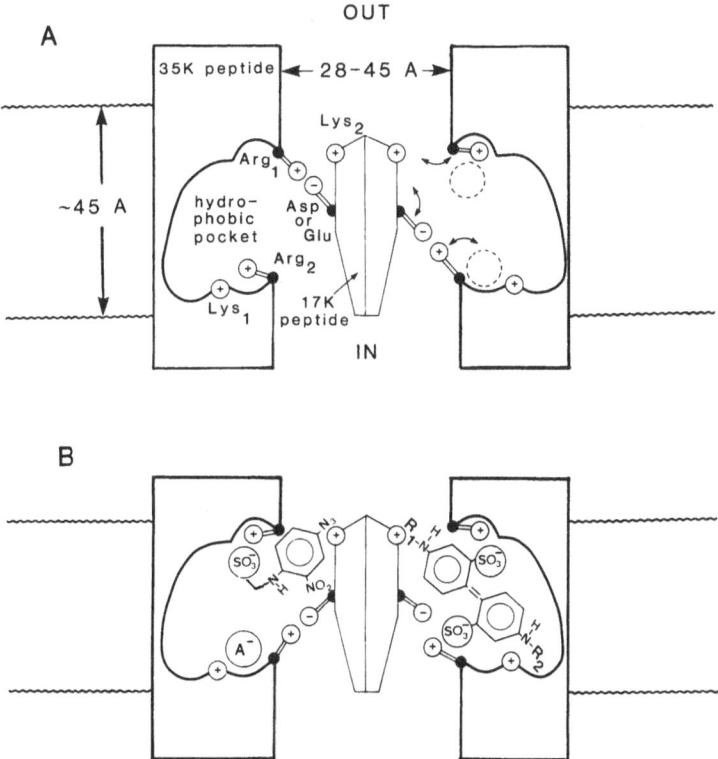

FIGURE 5. Hypothetical model of transport sites of a Band 3 dimer.

(A) Arrangement of residues at the transport site. Two anion sites are shown, only one of which is open at any time. Possible assignments of positive charged residues involved in transport have been made. Access to the transport site is controlled by an anionic gate (the Asp or Glu residue). When the outer transport site is closed, anions can still bind ("modifier"), but with lower affinity, and block movement of the gate. The two transport sites function independently but may be accessed via a common cavity between the subunits (see Macara and Cantley, 1981b).

(B) Binding of inhibitors at the transport sites of a Band 3 dimer. On the left, NAP-taurine is shown occupying the closed outer ("modifier") site. Substrate anions (A^-) can still bind. On the right, a stilbene disulfonate is shown occupying both the open inner (transport) and closed outer (modifier) sites.

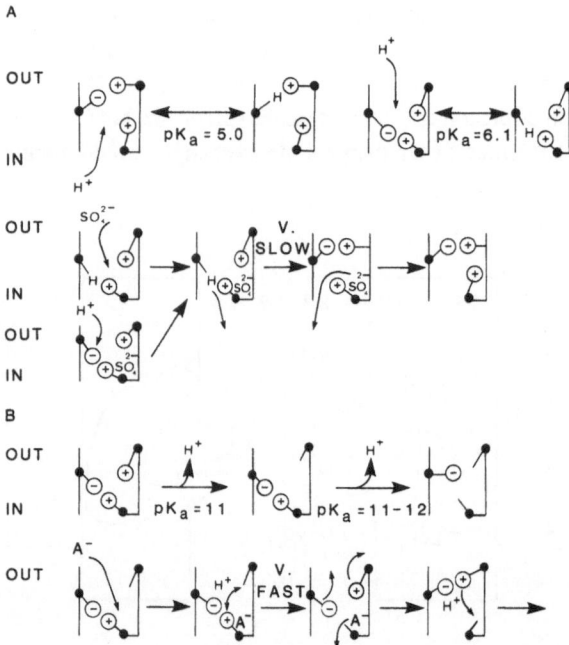

FIGURE 6. Hypothetical scheme for titration of the transport site residues of Band 3. (A) Titration of the anionic gate residue (Asp or Glu). (B) Titration of the arginine residues at the outer and inner anion binding sites. Access to both sites is possible from the external medium.

pletely blocked. Since the two pK_a values determined by experiment are assumed here to refer to the same residue in different conformations, the model predicts that imposition of a transmembrane pH gradient in the range of pH 5–6 would alter the proportion of inward- to outward-facing sites and hence alter the apparent affinities for substrate; conversely, imposing suitable substrate gradients to change the proportion of inward- to outward-facing sites would alter the apparent pK_as.

The model also predicts that at high pH the anion binding sites will be protonated, resulting in an inhibition of transport, and that the pK_a for the "transport site" arginine will be rather higher than for that of the "modifier site" because of the electrostatic interaction with the negatively charged gate residue when the sites are outwardly facing (Figure 6B). Anions bound at the substrate site would raise the pK_a even further, in agreement with the observations of Wieth and Bjerrum (1982). Stimulation of anion exchange on

membrane proteins other than Band 3 remain capable of DIDS-sensitive net sulfate flux (Wolosin, 1980). Differences in temperature and pH dependence and anion selectivity have been reported, but are to be expected because, even though the same protein is involved, the transport mechanisms must clearly differ.

There are at least two possible mechanisms for the net transport of anions by Band 3. One is "slippage" of the unloaded carrier: $E_{out} \to E_{in}$. The other is "tunneling," "transit," or "leakage" of anions with no change in the conformational state of the carrier (Fröhlich *et al.*, 1983; Kaplan *et al.*, 1980, 1982; Knauf and Law, 1980; Knauf *et al.*, 1983).

The slippage mechanism makes a very strong prediction, which is amenable to experimental test: Since the concentration of unloaded carrier ($E_{out} + E_{in}$) will decrease as the substrate concentration is raised, the rate of net anion transport should be inversely proportional to the substrate concentration. The leakage model, on the other hand, predicts that the net flux will be directly proportional. A rather weaker prediction is that, because the rate-limiting step in the slippage model is the $E_{out} \leftrightarrow E_{in}$ transition, the net flux should be independent of the identity of the anion, at anion concentrations well below the K_d for binding at the transport site. In the leakage model, however, the flux rate depends upon a permeability barrier, which will differ for different anions.

Comparisons of the net fluxes of different anions at low concentrations have not yet been reported. However, the concentration dependence of net chloride flux in the absence of a chloride gradient has been determined by two groups (Kaplan *et al.*, 1980; Knauf and Law, 1980), and both showed an increase in the flux with increasing chloride concentration, thus excluding the simple slippage model. An alternative slippage model is conceivable, in which binding of an anion to a separate site (the modifier site, for instance) accelerates the rate of translocation of the unloaded carrier, but the computed affinities for the transport and modifer site required to fit the available flux data are unrealistically low (I. G. Macara, unpublished calculations). The leakage model appears, therefore, to provide the best explanation of net anion transport.

It will be of interest to determine how the conformational state of Band 3 affects the anion permeability barrier. Initial results of Fröhlich *et al.* (1983) have shown that, in the presence of a large chloride gradient (very low extracellular concentration), such that most of Band 3 is in the E_{out} state, net flux is high but decreases rapidly with increasing extracellular chloride. This is to be expected as the ratio E_{out}/E_{in} decreases, if E_{in} has a lower net permeability to anions than does E_{out} (Knauf *et al.*, 1983), but further experiments are required to resolve the point.

4.2.3. Water Transport

Although the water and nonelectrolyte permeabilities of chicken erythrocytes are no greater than those of a lipid bilayer (Brahm and Wieth, 1977), there is some evidence from diffusion measurements for an aqueous pore in the human red cell (Solomon *et al.*, 1982, 1983). The data are best fit by assuming the presence of about 3×10^5 pores/cell, with a pore diameter of about 8 Å. The number of pores is fairly close to the number of Band 3 dimers in the red cell membrane, and Solomon and co-workers (1983) have suggested that an aqueous channel across the membrane is created at the interface between Band 3 subunits. A similar structure has been suggested to explain the negative cooperativity of stilbene disulfonate binding (Macara and Cantley, 1981a). One difficulty with assigning the aqueous pore to Band 3 is that stilbene disulfonates have no effect on water transport. Those sulfhydryl reagents that inhibit water transport, furthermore, do not affect anion exchange (Knauf, 1979). Surprisingly, however, they do appear to react with Band 3. ^{14}C-labeled 5,5'-dithiobis(2-nitrobenzoic acid), for instance, reacts almost exclusively with a 95-Kd membrane protein when incubated with *N*-ethylmaleimide- and iodoacetamide-protected red cells (Brown *et al.*, 1975). Unfortunately, neither the stoichiometry of reaction nor the effect of DIDS were determined, so that the identification with Band 3 remains inconclusive. Another water transport inhibitor, para-(chloromercuri)benzene sulfonic acid (PCMBS), has been investigated more carefully, however, and shown to bind to specific extracellular sites in resealed ghosts with a stoichiometry of 1–2 $\times 10^6$/ghost (Rao, 1978; Solomon *et al.*, 1982). Gel electrophoresis using [203Hg]-PCMBS indicated that the inhibitor reacts almost exclusively with Band 3. Although stilbene disulfonates have no effect on PCMBS binding (Rao, 1978), it appears that PCMBS does decrease the rate of the conformational change that locks stilbene disulfonates into their binding site on Band 3 (Solomon *et al.*, 1982). It has been suggested, therefore, that PCMBS reacts at a site (possibly the cryptic SH group in the 17-Kd peptide) close to the extracellular surface of the membrane so that it can block access to the anion inhibitor site. An allosteric mechanism has not been excluded, however.

Thus considerable evidence now exists as to the identity of the aqueous pore and Band 3, but proof remains elusive and will clearly require the use of purified Band 3 reconstituted into phospholipid vesicles.

5. BIOSYNTHESIS OF BAND 3

Secreted proteins, and integral membrane proteins like glycophorin HLA and VSV-G that only span the bilayer once ("type 1" membrane proteins), are synthesized with an extra, hydrophobic, "signal" peptide of about 23 residues at the N terminus. The peptide controls the insertion of the nascent protein into the membrane and is usually cleaved during insertion by a specific, membrane-bound protease. This mechanism naturally results in the extrusion of the N terminus on the extracellular side of the membrane. The Band 3 polypeptide, however, crosses the bilayer at least three times, and the N terminus remains cytoplasmic. There is considerable interest, therefore, in the mechanism by which Band 3 and other integral membrane proteins of similar type (the Na^+,K^+-ATPase and Ca^{2+}-ATPase, for instance) are inserted.

Although the technology for studying the biogenesis of membrane proteins was developed several years ago, its application to Band 3 has been delayed by the difficulty in obtaining sufficient active and intact mRNA. Recently, however, Lodish and co-workers have described a cell-free system for Band 3 synthesis using RNA from the spleen cells of anemic mice (Braell and Lodish, 1981, 1982). They found that, like most other membrane proteins, Band 3 is cotranslationally inserted into the membranes of the endoplasmic reticulum, but it does not possess a cleavable N-terminal leader sequence. Immediately after synthesis, the protein appears in its mature configuration, with the N-terminal domain exposed to the cytoplasm and with a high-mannose oligosaccharide attached to the external surface of the membrane-bound domain. This precursor oligosaccharide is presumably modified in the Golgi bodies en route to the cell surface, as is the case with type 1 membrane proteins. In synchronized translation studies it was found that insertion and glycosylation of the nascent Band 3 polypeptide could occur surprisingly late in synthesis. All other co-translationally inserted membrane proteins studied to date only insert correctly if microsomes are added before the peptide is more than ~100 residues long. With Band 3, however, microsomes could be added as late as 35–40 min after initiation of synthesis, which corresponds to a length of about 450–500 residues (Braell and Lodish, 1982). The signal sequence for membrane insertion therefore appears to be internal and to be localized near the connection between the cytoplasmic and membrane-bound domains.

An interesting speculation, arising from this result, is that Band 3 evolved from two distinct proteins, one of which was originally cytoplasmic, while the other was an integral membrane protein with a normal N-terminal signal sequence. Conflation of the two into a single polypeptide then resulted in the internalization of the sequence. Although such a hypothesis is difficult to prove, it will be of great interest to determine if the domain structure of Band 3 is represented by separate exons in the chromosome and if the signal sequence is actually located at the N terminus of the membrane domain exon.

6. CONCLUSION

A clear picture of Band 3 as a two-domain, multifunctional protein has emerged within the last few years. Rapid progress has been made in elucidating the kinetic mechanism of anion exchange and the structure of the active site and in our understanding of the multifarious interactions of Band 3 with other proteins. This progress can only accelerate as new techniques are utilized. In particular, we can expect molecular genetics to provide the complete primary sequence of the protein within a year or so. Various computer programs already exist for calculating the probable folding patterns for membrane proteins from their sequences, and these, together with further chemical modification studies, will allow a much clearer picture of the transport site of Band 3 to be drawn. However, even if the complete three-dimensional crystal structure of the protein is eventually known, sophisticated biochemical and kinetic studies will still be required in order to elucidate the integration of the various functions of Band 3 with the metabolism of the red cell.

ACKNOWLEDGMENTS. We wish to thank Guido Guidotti, Ted Steck, Phil Knauf, Herman Passow, Mike Jennings, Jim Salhany, Otto Wieth, Aser Rothstein, and many others for providing unpublished data and for helpful discussion. Financial support during the period in which this review was written was provided by a Charles King Trust fellowship (to I.G.M.) and by Grant GM 26199 from the National Institutes of Health (to L.C.C.).

REFERENCES

Appell, K. C., and Low, P. S., 1981, Partial structural characterization of the cytoplasmic domain of the erythrocyte membrane protein, Band 3, *J. Biol. Chem.* **256**:11104–11111.

Appell, K. C., and Low, P. S., 1982, Evaluation of structural independence of membrane-spanning and cytoplasmic domains of Band 3, *Biochemistry* **21**:2151–2157.

Baskin, G. S., and Langdon, R. G., 1981, A spectrin-dependent ATPase of the human erythrocyte membrane, *J. Biol. Chem.* **256**:5428–5435.

Barzilay, M., Ship, S., and Cabantchik, Z. I., 1979, Anion transport in red blood cells. I. Chemical properties of anion recognition sites as revealed by structure–activity relationships of aromatic sulfonic acids, *Membr. Biochem.* **2**:227–254.

Bayley, H., and Knowles, J. R., 1979, Photogenerated reagents for membranes: Selective labeling of intrinsic membrane proteins in the human erythrocyte membrane, *Biochemistry* **19**:3883–3892.

Bennett, V., and Stenbuck, P. J., 1979, The membrane attachment protein for spectrin is associated with Band 3 in human erythrocyte membranes, *Nature* **280**:468–473.

Bennett, V., and Stenbuck, P. J., 1980, Association between ankyrin and the cytoplasmic domain of Band 3 isolated from the human erythorycte membrane, *J. Biol. Chem.* **255**:6424–6432.

Beth, A. H., Balasubramanian, K., Wildes, R. T., Venkatasamie, S. D., Robinson, B. H., Dalton, L. R., Pearson, D. E., and Park, J. H., 1981, Structural and motional changes in glyceraldehyde-3-phosphate dehydrogenase upon binding to the Band-3 protein of the erythrocyte membrane examined with [^{15}N, ^{3}H] maleimide spin label and electron paramagnetic resonance, *Proc. Natl. Acad. Sci. USA* **78**:4955–4959.

Bjerrum, P. J., Borders, C. L., Jr., and Wieth, J. O., 1982a, Arginyl residues at anion transport sites: Modification, location and effect on DIDS-binding in human red cells, *Fed Proc.* **41**:975 (abstr.).

Bjerrum, P. J., Wieth, J. O., and Borders, C. L., Jr., 1982b, An arginyl residue is essential for anion transport in red blood cells, *Acta Physiol. Scand.* **114**:4A (abstr.).

Braell, W. A., and Lodish, H. F., 1981, Biosynthesis of the erythrocyte anion transport protein, *J. Biol. Chem.* **256**:11337–11344.

Braell, W. A., and Lodish, H. F., 1982, The erythrocyte anion transport protein is cotranslationally inserted into microsomes, *Cell* **28**:23–31.

Brahm, J., 1977, Temperature-dependent changes of chloride transport kinetics in human red blood cells, *J. Gen. Physiol.* **70**:283–306.

Brahm, J., and Wieth, J. O., 1977, Separate pathways for urea and water, and for chloride in chicken erythrocytes, *Nature* **266**:727–749.

Branton, D., Cohen, C. M., and Tyler, J., 1981, Interactions of cytoskeletal proteins on the human erythrocyte membrane, *Cell* **24**:24–32.

Brown, P. A., Feinstein, M. B., and Sha'afi, R. I., 1975, Membrane proteins related to water transport in human erythrocytes, *Nature* **254**:523–525.

Cabantchik, Z. I., Knauf, P. A., Ostwald, T., Markus, H., Davidson, L., Breuer, L. S., and Rothstein, A., 1976, The interactions of an anionic photoreactive probe with the anion transport system of the human red blood cell, *Biochim. Biophys. Acta* **455**:526–537.

Cabantchik, Z. I., Knauf, P. A., and Rothstein, A., 1978, The anion transport system of the red blood cell. The role of membrane protein evaluated by the use of "probes," *Biochim. Biophys. Acta* **515**:239–302.

Cabantchik, Z. I., Volsky, D. J., Ginsburg, H., and Loyter, A., 1980, Reconstitution of the erythrocyte anion transort system: *In vitro* and *in vivo* approaches, *Ann. N. Y. Acad. Sci.* **341**:444–454.

Cherry, R. J., and Nigg, E. A., 1980, Molecular interactions involving Band 3: Information from rotational diffusion measurements, in: *Membrane Transport in Erythrocytes* (U. V. Lassen, H. H. Ussing, and J. O. Wieth, eds.), Alfred Benzon Symposium 14, Munksgaard, Copenhagen, pp. 130–138.

Cousin, J. L., and Motais, R., 1979, Inhibition of anion permeability by amphiphilic compounds in human red cell: Evidence for an interaction of niflumic acid with the Band 3 protein, *J. Memb. Biol.* **46**:125–153.

Dalmark, M., 1976, Effects of halides and bicarbonate on chloride transport in human red blood cells, *J. Gen. Physiol.* **67**:223–234.

Dix, J. A., Verkman, A. S., Solomon, A. K., and Cantley, L. C., 1979, Human erythrocyte anion exchange site characterized using a fluorescent probe, *Nature* **282**:520–522.

Dorst, H. -J., and Schubert, D., 1979, Self-association of Band 3-protein from human erythrocyte membranes in aqueous solutions, *Hoppe-Seyler's Z. Physiol. Chem.* **360**:1605–1618.

Drickamer, L. K., 1976, Fragmentation of the 95,000 dalton transmembrane polypeptide in human erythrocyte membranes. Arrangement of the fragments in the Lipid bilayer, *J. Biol. Chem.* **251**:5115–5123.

Drickamer, L. K., 1977, Fragmentation of the Band 3 polypeptide from human erythrocyte membranes. Identification of regions likely to interact with the lipid bilayer, *J. Biol. Chem.* **252**:6906–6917.

Drickamer, L. K., 1978, Orientation of the Band 3 polypeptide from human erythrocyte membranes. Identification of the NH_2-terminal sequence and site of carbohydrate attachment, *J. Biol. Chem.* **253**:7242–7248.

DuPre, A. M., and Rothstein, A., 1981, Inhibition of anion transport associated with chymotryptic cleavages of red blood cell Band 3 protein, *Biochim. Biophys. Acta* **646**:471–478.

Eisinger, J., Flores, J., and Salhany, J. M., 1982, Association of cytosol-hemoglobin with the membrane in intact erythrocytes, *Proc. Natl. Acad. Sci. USA* **79**:408–412.

Fairbanks, G., Steck, T. L., and Wallach, D. F. H., 1971, Electrophoretic analysis of the major polypeptides of the human erythrocyte membrane, *Biochemistry* **10**:2606–2617.

Fossel, E. T., and Solomon, A. K., 1977, Membrane mediated link between ion transport and metabolism in human red cells, *Biochim. Biophys. Acta* **464**:82–92.

Fossel, E. T., and Solomon, A. K., 1979, Effect of the sodium/potassium ratio on glyceraldehyde-3-phosphate dehydrogenase interaction with red cell vesicles, *Biochim. Biophys. Acta* **553**:142–153.

Freitag, C., and Clark, S., 1981, Reversible methylation of cytoskeletal and membrane proteins in intact human erythrocytes, *J. Biol. Chem.* **256**:6102–6108.

Fröhlich, O., and Gunn, R. B., 1982, Mutual interactions of reversible inhibitors on the red cell anion transporter, *Biophys. J.* **37**:213a.

Fröhlich, O., Leibson, C., and Gunn, R. B., 1983, Chloride net flux from intact erythrocytes under slippage conditions. Evidence for a positive charge on the anion binding/transport site, *J. Gen. Physiol.* (in press).

Fukuda, M., Eshdat, Y., Tarone, G., and Marchesi, V. T., 1978, Isolation and characterization of peptides derived from the cytoplasmic segment of Band 3, the predominant intrinsic membrane protein of the human erythrocyte, *J. Biol. Chem.* **253**:2419–2428.

Funder, J., and Wieth, J., 1976, Chloride transport in human erythrocytes and ghosts: A quantitative comparison, *J. Physiol.* **262**:679–698.

Ginsburg, H., O'Connor, S. E., and Grisham, C. M., 1981, Evidence from electron paramagnetic resonance for function-related conformation changes in the anion-transport protein of human erythrocytes, *Eur. J. Biochem.* **114**:533–538.

Golan, D. E., and Veatch, W., 1980, Lateral mobility of Band 3 in the human erythrocyte membrane studied by fluorescence photobleaching recovery: Evidence for control by cytoskeletal interactions, *Proc. Natl. Acad. Sci. USA* **77**:2537–2541.

Golan, D. E., and Veatch, W., 1982, Lateral mobility of Band 3 in the human erythrocyte membrane: Control by ankyrin-mediated interactions, *Biophys. J.* **37**:177a.

Goldberg, M. A., Lalos, A. T., and Bunn, H. F., 1981, The effect of erythrocyte membrane preparations on the polymerization of sicle hemoglobin, *J. Biol. Chem.* **256**:193–197.

Grinstein, S., Ship, S., and Rothstein, A., 1978, Anion transport in relation to proteolytic dissection of Band 3 protein, *Biochim. Biophys. Acta* **507**:294–304.

Grinstein, S., McCulloch, L., and Rothstein, A., 1979, Transmembrane effects of irreversible inhibitors of anion transport in red blood cells. Evidence for mobile transport sites, *J. Gen. Physiol.* **73**:493–514.

Guidotti, G., 1980, The structure of the Band 3 polypeptide, in: *Membrane Transport in Erythrocytes* (U. V. Lassen, H. H. Ussing, and J. O. Wieth, eds.), Alfred Benzon Symposium 14, Munksgaard, Copenhagen, pp. 300–308.

Gunn, R. B., 1978, Considerations of the titratable carrier model for sulfate transport in human red blood cells, *Membr. Transp. Processes* **1**:61–77.

Gunn, R. B., and Fröhlich, O., 1979, Asymmetry in the mechanism for anion exchange in human red blood cell membranes. Evidence for reciprocating sites that react with one transported anion at a time, *J. Gen. Physiol.* **74**:351–374.

Hargreaves, W. R., Giedd, K. N., Verkleij, A., and Branton, D., 1980, Reassociation of ankyrin with Band 3 in erythrocyte membranes and in lipid vesicles, *J. Biol. Chem.* **255**:11965–11972.

Harper, P. A., and Knauf, P. A., 1979, Comparison of chloride transport in mouse erythrocytes and Friend virus-transformed erythroleukemia cells, *J. Cell Physiol.* **98**:347–358.

Higashi, T., Richards, C. S., and Uyeda, K., 1979, The interaction of phosphofructokinase with erythrocyte membranes, *J. Biol. Chem.* **254**:9542–9550.

Hoffman, J. F., and Proverbio, F., 1974, Membrane ATP and the functional organization of the red cell Na^+-K^+-pump, *Ann. N. Y. Acad. Sci.* **242**:459–460.

Jenkins, R. E., and Tanner, M. J. A., 1977, the structure of the major protein of the human erythrocyte membrane. Characterization of the intact protein and major fragments, *Biochem. J.* **161**:139–147.

Jennings, M. L., 1976, Proton fluxes associated with erythrocyte membrane anion exchange, *J. Membr. Biol.* **28**:187–205.

Jennings, M. L., 1980, Apparent "recruitment" of SO_4 transport sites by the Cl gradient across the human erythrocyte membrane, in: *Membrane Transport in Erythrocytes* (U. V. Lassen, H. H. Ussing, and J. O. Wieth, eds.), Alfred Benzon Symposium 14, Munksgaard, Copenhagen, pp. 450–463.

Jennings, M. L., 1982a, Reductive methylation of the two H_2DIDS-binding lysine residues of Band 3, the human erythrocyte anion transport protein, *J. Biol. Chem.* **257**:7554–7559.

Jennings, M. L., 1982b, Reductive methylation of two H_2DIDS-binding lysine residues in Band 3, the human erythrocyte anion transport protein, *Biophys. J.* **37**:177a.

Jennings, M. L., 1982c, Stoichiometry of a half-turnover of Band 3, the chloride transport protein of human erythrocytes, *J. Gen. Physiol* **79**:169–185.

Jennings, M. L., and Adams, M. F., 1981, Modification by papain of the structure and function of Band 3, the erythrocyte anion transport protein, *Biochemistry* **20:**7118–7122.

Jennings, M. L., and Passow, H., 1979, Anion transport across the erythrocyte membrane, *in situ* proteolysis of Band 3 protein, and cross-linking of proteolytic fragments by 4,4'-diisothiocyano-dihydrostilbene-2,2'-disulfonate, *Biochim. Biophys. Acta* **554:**498–519.

Kant, J. A., and Steck, T. L., 1974, Preparation of impermeable ghosts and inside-out vesicles from human erythrocyte membranes, *Methods Enzymol.* **31:**172–180.

Kaplan, J. H., Scorah, K., Fasold, H. and Passow, H., 1976, Sidedness of the inhibitory action of disulfonic acids on chloride equilibrium exchange and net transport across the human erythrocyte membrane, *FEBS Lett.* **62:**182–185.

Kaplan, J. H., Pring, M., and Passow, H., 1980, Concentration dependence of chloride movements that contribute to the conductance of the red cell membrane, in: *Membrane Transport in Erythrocytes* (U. V. Lassen, H. H. Ussing, and J. O. Wieth, eds.), Alfred Benzon Symposium 14, Munksgaard, Copenhagen, pp. 494–497.

Kaplan, J. H., Pring, M., and Passow, H., 1982, Band-3-mediated diffusive anion flow across the red blood cell membrane, *Fed. Proc.* **41:**975 (abstr.).

Karadsheh, N. S., and Uyeda, K., 1977, Changes in allosteric properties of phosphofructokinase bound to erythrocyte membranes, *J. Biol. Chem.* **252:**7418–7420.

Kempf, C., Brock, C., Sigrist, H., Tanner, M. J. A., and Zahler, P., 1981, Interaction of phenylisothiocyanate with human erythrocyte Band 3 protein. II. Topology of phenyliso-thiocyanate binding sites and influence of p-sulfophenylisothiocyanate on phenylisothio-cyanate modification, *Biochim. Biophys. Acta* **641:**88–98.

Kleinfeld, A. M., Lukacovic, M., Matayoshi, E. D., and Holloway, P., 1982, Conformation of membrane proteins determined from the spatial distribution of tryptophan, *Biophys. J.* **37:**146a.

Kleinfeld, A. M., Matayoshi, D. E., and Solomon, A. K., 1980, Use of Band 3 vesicles from human erythrocytes to study protein structural changes associated with anion transport, *Fed. Proc.* **39:**1714 (abstr.).

Kliman, H. J., and Steck, T. L., 1980, Association of glyceraldehyde-3-phosphate dehydrogenase with the human red cell membrane. A kinetic analysis, *J. Biol. Chem.* **255:**6314–6321.

Knauf, P. A., 1979, Erythrocyte anion exchange and the Band 3 protein: Transport kinetics and molecular structure, *Curr. Topics Membr. Transp.* **912:**249–363.

Knauf, P. A., and Law, F.-Y., 1980, Relationship of net anion flow to the anion exchange system, in: *Membrane Transport in Erythrocytes* (U. V. Lassen, H. H. Ussing, and J. O. Wieth, eds.), Alfred Benzon Symposium 14, Munksgaard, Copenhagen, pp. 488–493.

Knauf, P. A., and Mann, N., 1982, Use of niflumic acid (NA) to probe the asymmetry of the human erythrocyte anion exchange system, *Fed. Proc.* **41:**975 (abstr.).

Knauf, P. A., Breuer, W., McCulloch, L., and Rothstein, A., 1978a, N-(4-azido-2-nitrophenyl)-2-aminoethylsulfonate (NAP-taurine) as a photoaffinity probe for identifying membrane components containing the modifier site of the human red blood cell anion exchange system, *J. Gen. Physiol.* **72:**631–649.

Knauf, P. A., Fuhrmann, G. F., Rothstein, S., and Rothstein, A., 1977, The relationship between anion exchange and net anion flow across the human red blood cell membrane, *J. Gen. Physiol* **69:**363–386.

Knauf, P. A., Law, F. -Y., and Marchant, P. M., 1983, Relationship of net chloride flow across the human erythrocyte membrane to the anion exchange mechanism, *J. Gen. Physiol.* (in press).

Knauf, P. A., Ship, S., Breuer, W., McCulloch, L., and Rothstein, A., 1978b, Asymmetry of the red cell anion exchange system: Different mechanisms of reversible inhibition by N-(4-

azido-2-nitrophenyl)-2-aminoethyl-sulfonate (NAP-taurine) at the inside and outside of the membrane, *J. Gen. Physiol.* **72:**607–630.

Knauf, P. A., Tarshis, T., Grinstein, S., and Furuya, W., 1980, Spontaneous and induced asymmetry of the human erythrocyte anion exchange system as detected by chemical probes, in: *Membrane Transport in Erythrocytes* (U. V. Lassen, H. H. Ussing, and J. O. Wieth, eds.), Alfred Benzon Symposium 14, Munksgaard, Copenhagen, pp. 389–403.

Köhne, W., Huest, C. W. M., and Deuticke, B., 1981, Mediated transport of anions in Band 3-phospholipid vesicles, *Biochim. Biophys. Acta* **664:**108–120.

Kotaki, A., Naoi, M., and Yagi, K., 1971, A diaminostilbene dye as a hydrophobic probe for proteins, *Biochim. Biophys. Acta* **229:**547–556.

Ku, C. -P., Jennings, M. L., and Passow, H., 1979, A comparison of the inhibitory potency of reversibly acting inhibitors of anion transport on chloride and sulfate movements across the human red cell membrane, *Biochim. Biophys. Acta* **553:**132–144.

Legrum, B., Fasold, H., and Passow, H., 1980, Enhancement of anion equilibrium exchange by densylation of the red blood cell membrane, *Hoppe-Seyler's Z. Physiol. Chem.* **361:**1573–1590.

Lepke, S., Fusold, H., Pring, M., and Passow, H., 1976, A study of the relationship between inhibition of anion exchange and binding to the red blood cell membrane of 4,4'-diisothio-cyano-stilbene-2,2'-disulfonic acid (DIDS) and of its dihydro derivative (H_2DIDS), *J. Membr. Biol.* **29:**147–177.

Lukacovic, M. F., Feinstein, M. B., Sha'afi, R. I., and Perrie, S., 1981, Purification of stabilized Band 3 protein of the human erythrocyte membrane and its reconstitution into liposomes, *Biochemistry* **20:**3145–3151.

Macara, I. G., and Cantley, L. C., 1981a, Interactions between transport inhibitors at the anion binding sites of the Band 3 dimer, *Biochemistry* **20:**5095–5105.

Macara, I. G., and Cantley, L. C., 1981b, Mechanism of anion exchange across the red cell membrane by Band 3: Interactions between stilbene-disulfonate and NAP-taurine binding sites, *Biochemistry* **20:**5695–5701.

Macara, I. G., Kuo, S., and Cantley, 1982, Inhibitors of anion exchange induce a transconformational change in Band 3, *Biophys J.* **37:**143a.

Macara, I. G., Kuo, S., and Cantley, L. C., 1983, Evidence that inhibitors of anion exchange induce a transmembrane conformational change in Band 3, *J. Biol. Chem.* **258** (in press).

Markowitz, S., and Marchesi, V. T., 1981, The carboxyl-terminal domain of human erythrocyte Band 3. Description, isolation, and location in the bilayer, *J. Biol. Chem.* **256:**6463–6468.

Mercer, R. W., and Dunham, P. B., 1982, Membrane-bound ATP fuels the Na/K pump. Studies on membrane-bound glycolytic enzymes on inside-out vesicles from human red cell membranes, *J. Gen. Physiol.* **78:**547–568.

Milanick, M. A., and Gunn, R. B., 1981, The selectivity of the external anion exchange mechanism of human red blood cells, *Biophys. J.* **33:**47a.

Milanick, M. A., and Gunn, R. B., 1982a, Proton–sulfate co-transport. Mechanism of H^+ and sulfate addition to the chloride transporter of human red blood cells, *J. Gen. Physiol.* **79:**87–113.

Milanick, M. A., and Gunn, R. B., 1982b, Interactions between external protons and the anion transporter of human erythrocytes, *Biophys. J.* **37:**213a.

Murthy, S. N. P., Liu, T., Köhler, H., and Steck, T. L., 1981a, The aldolase binding site of the human erythrocyte membrane. Primary structure of the amino-terminal decapeptide of Band 3, *J. Supramol. Struct. Suppl.* **5:**125.

Murthy, P. S. N., Liu, T., Kaul, R. K., Köhler, H., and Steck, T. L., 1981b, The aldolase-binding site of the human erythrocyte membrane is at the NH_2 terminus of Band 3, *J Biol. Chem.* **256:**11203–11208.

Nakashima, H., Nakagawa, Y., and Makino, S., 1981, Detection of the associated state of membrane proteins by polyacrylamide gradient gel electrophoresis with non-denaturing detergents. Application to Band 3 protein from erythrocyte membranes, *Biochim. Biophys. Acta* **643**:509–518.

Nigg, E., and Cherry, R. J., 1979, Dimeric association of Band 3 in the erythrocyte membrane is demonstrated by protein diffusion measurements, *Nature* **277**:493–494.

Nigg, E. A., and Cherry, R. J., 1980, Anchorage of a Band 3 population at the erythrocyte cytoplasmic membrane surface: Protein rotational diffusion measurements, *Proc. Natl. Acad. Sci. USA* **77**:4702–4706.

Pappert, G., and Schubert, D., 1982, Self-association of Band 3 protein from erythrocyte membranes in solutions of a non-ionic detergent, Ammonyx-LO, *Protides Biol. Fluids, Proc. Colloq.* **29**:117–121.

Passow, H., Fasold, H., Gartner, M., Legrum, B., Ruffing, W., and Zaki, L., 1980a, Anion transport across the red blood cell membrane and the conformation of the protein in Band 3, *Ann. N. Y. Acad. Sci.* **341**:361–383.

Passow, H., Kampmann, L., Fasold, H., Jennings, M., and Lepke, S., 1980b, Mediation of anion transport across the red blood cell membrane by means of conformational changes of the Band 3 protein, in: *Membrane Transport in Erythrocytes* (U. V. Lassen, H. H. Ussing, and J. O. Wieth, eds.), Alfred Benzon Symposium 14, Munksgaard, Copenhagen, pp. 345–367.

Premachandra, B. R., and Mentzer, W. C., 1980, Studies on the high affinity binding site for normal (AA) and sickle cell (SS) hemoglobins on AA and SS inside-out erythrocyte membrane vesicles, *Fed. Proc.* **39**:1916 (abstr.).

Ramjeesingh, M., and Rothstein, A., 1983, The location of a chymotrypsin cleavage site and of other sites in the primary structure of the 17,000 dalton transmembrane segment of Band 3, the anion transport protein of the red cell, *Membr. Biochem.* (in press).

Ramjeesingh, M., Gaarn, A., and Rothstein, A., 1980a, The location of a disulfonic stilbene binding site in Band 3, the anion transport protein of the red blood cell membrane, *Biochim. Biophys. Acta* **599**:127–139.

Ramjeesingh, M., Grinstein, S., and Rothstein, A., 1980b, Intrinsic segments of Band 3 that are associated with anion transport across red blood cell membranes, *J. Membr. Biol.* **57**:95–102.

Ramjeesingh, M., Gaarn, A., and Rothstein, A., 1982, The sulfhydryl groups of the 35,000 dalton C-terminal segment of Band 3 are located in a 9,000 dalton fragment produced by chymotrypsin treatment of red cell ghosts, *J. Bioenerg. Biomembr.* **13**:411–423.

Rao, A., 1978, Reactice sulfhydryl groups of Band 3 of red cell membranes, Ph.D. Thesis, Department of Biochemistry and Molecular Biology, Harvard University, Cambridge, Massachusetts.

Rao, A., 1979, Disposition of the Band 3 polypeptide in the human erythrocyte membrane. The relative sulfhydryl groups, *J. Biol. Chem.* **254**:3503–3511.

Rao, A., and Reithmeier, R. A. F., 1979, Reactive sulfhydryl groups of the Band 3 polypeptide from human erythrocyte membranes. Locations in the primary structure, *J. Biol. Chem.* **254**:6144–6150.

Rao, A., Martin, P., Reithmeier, R. A. F., and Cantley, L. C., 1979, Location of the stilbene disulfonate binding site of the human erythrocyte anion-exchange system by resonance energy transfer, *Biochemistry* **18**:4505–4516.

Reithmeier, R. A. F., 1979, Fragmentation of the Band 3 polypeptide from human erythrocyte membranes. Size and detergent binding of the membrane-associated domain, *J. Biol. Chem.* **254**:3054–3060.

Reithmeier, R. A. F., and Rao, A., 1979, Reactive sulfhydryl groups of the Band 3 polypeptide from human erythrocyte membranes. Identification of the sulfhydryl groups involved in Cu^{2+}-o-phenanthroline cross-linking, *J. Biol. Chem.* **254**:6151–6155.

Ross, A. H., and McConnell, H. M., 1978, Reconstitution of the erythrocyte anion channel, *J. Biol. Chem.* **253**:4777–4782.

Rothstein, A., Cabantchik, Z. I., and Knauf, P., 1976, Mechanism of anion transport in red blood cells: Role of membrane proteins, *Fed. Proc.* **35**:3–10.

Salhany, J. M., and Gaines, E. D., 1981, Steady state kinetics of erythrocyte anion exchange. Evidence for site–site interactions, *J. Biol. Chem.* **256**:11080–11085.

Salhany, J. M., and Shaklai, N., 1979, Functional properties of human hemoglobin bound to the erythrocyte membrane, *Biochemistry* **18**:893–899.

Salhany, J. M., Cordes, K. A., and Gaines, E. D., 1980, Light-scattering measurements of hemoglobin binding to the erythrocyte membrane. Evidence for transmembrane effects related to a disulfonic stilbene binding to Band 3, *Biochemistry* **19**:1447–1454.

Sayare, M., and Fikiet, M., 1981, Cross-linking of hemoglobin to the cytoplasmic surface of human erythrocyte membranes. Identification of Band 3 as a site for hemoglobin binding in Cu^{2+}-o-phenanthroline catalyzed cross-linking, *Proc. Natl. Acad. Sci. USA* **256**: 13152–13158.

Schneider, A. B., Dean, A., and Schechter, A. N., 1980, Binding of hemoglobin to erythrocyte ghosts measured by a competitive binding (radioreceptor) assay, *Fed. Proc.* **39**:1917 (abstr.).

Schnell, K. F., Gerbardt, S., and Schoppe-Fredenburg, A., 1977, Kinetic characteristics of the sulfate self-exchange in human red blood cells and red blood cell ghosts, *J. Membr. Biol.* **30**:319–350.

Shaklai, N., Yguerabide, J., and Ranney, H. M., 1977a, Interaction of hemoglobin with red blood cell membranes as shown by a fluorescent chromaphore, *Biochemistry* **16**:5585–5592.

Shaklai, N., Yguerabide, J., and Ranney, H. M., 1977b, Classification and localization of hemoglobin binding sites on the red blood cell membrane, *Biochemistry* **16**:5593–5597.

Shaklai, N., Sharma, V. S., and Ranney, H. M., 1981, Interaction of sickle cell hemoglobin with erythrocyte membranes, *Proc. Natl. Acad. Sci. USA* **78**:65–68.

Ship, S., Shami, Y., Breuer, W., and Rothstein, A., 1977, Synthesis of tritiated 4,4'-diiso-thiocyano-2,2'-stilbene disulfonic acid [(^{3}H)$_2$DIDS] and its covalent reaction with sites related to anion transport in red blood cells, *J. Membr. Biol.* **33**:311–324.

Sigrist, H., Kempff, C., and Zahler, P., 1980, Interaction of phenylisothiocyanate with human erythrocyte Band 3. 1. Covalent modification and inhibition of phosphate transport, *Biochim. Biophys. Acta* **597**:137–144.

Solomon, A. K., Chasan, B., Dix, J. A., Lukacovic, M. F., Toon, M. R., and Verkman, A. S., 1982, The aqueous pore in the red cell membrane: Band 3 as a channel for anions, cations, nonelectrolytes, and water, *Biophys. J.* **37**:215a.

Solomon, A. K., Chason, B., Dix, J. A., Lukacovic, M. F., Toon, M. R., and Verkman, A. S., 1983, The aqueous pore in the red cell membrane: Band 3 as a channel for anions, cations, non-electrolytes, and water, *Ann. N.Y. Acad. Sci.* (in press).

Steck, T. L., 1974, The organization of proteins in the human red cell membrane, *J. Cell Biol.* **62**:1–19.

Steck, T. L., 1978, The Band 3 protein of the human red cell membrane: A review, *J. Supramol. Struct.* **8**:311–324.

Steck, T. L., Ramos, B., and Stapazon, E., 1976, Proteolytic dissection of Band 3, the predominant transmembrane polypeptide of the human erythrocyte membrane, *Biochemistry* **15**:1154–1161.

Steck, T. L., Koziarz, J. J., Singh, M. K., Reddy, R., and Köhler, H., 1978, Preparation and analysis of seven major, topographically defined fragments of Band 3, the predominant transmembrane polypeptide of human erythrocyte membranes, *Biochemistry* **17**:1216–1222.

Strapazon, E., and Steck, T. L., 1976, Binding of rabbit muscle aldolase to Band 3, the predominant polypeptide of the human erythrocyte membrane, *Biochemistry* **15**:1421–1424.

Strapazon, E., and Steck, T. L., 1977, Interaction of the aldolase and the membrane of human erythrocytes, *Biochemistry* **16**:2966–2971.

Stryer, L., 1978, Fluorescence energy transfer as a spectroscopic ruler, *Annu. Rev. Biochem.* **47**:819–846.

Tanner, M. J. A., 1979, Isolation of integral membrane proteins and criteria for identifying carrier proteins, *Curr. Top. Membr. Transp.* **12**:1–51.

Terwilliger, T. C., and Clark, S., 1981, Methylation of membrane proteins in human erythrocytes. Identification and characterization of polypeptides methylated in lysed cells, *J. Biol. Chem.* **256**:3067–3076.

Tsai, I.-O., Murthy, S. N. P., and Steck, T. L., 1982, Effect of red cell membrane binding on the catalytic activity of glyceraldehyde-3-phosphate dehydrogenase, *J. Biol. Chem.* **257**:1438–1442.

Tsuji, T., Irimura, T., and Osawa, T., 1980, The carbohydrate moiety of Band 3 glycoprotein of human erythrocyte membranes, *Biochem. J.* **187**:677–686.

Tsuji, T., Irimura, T., and Osawa, T., 1981, The carbohydrate moiety of Band 3 glycoprotein of human erythrocyte membranes. Structure of lower molecular weight oligosaccharides, *J. Biol. Chem.* **256**:10497–10502.

Volsky, D. J., Cabantchik, Z. I., Beigel, M., and Loyter, A., 1979, Implantation of the isolated human erythrocyte anion channel into plasma membranes of Friend erythroleukemia cells by use of Sendai virus envelopes, *Proc. Natl. Acad. Sci. USA* **75**:5440–5444.

Weinstein, R. S., Rhodadad, J. K., and Steck, T. L., 1980, The Band 3 protein intramembrane particle of the human red blood cell, in: *Membrane Transport in Erythrocytes* (U. V. Lassen, H. H. Ussing, and J. O. Wieth, eds.), Alfred Benzon Symposium 14, Munksgaard, Copenhagen, pp. 35–48.

Weinstein, R. S., Khodadad, J. K., and Steck, T. L., 1978, Fine structure of the Band 3 protein in human red cell membranes: Freeze–fracture studies, *J. Supramol. Struct.* **8**:325–335.

Wells, E., and Findlay, J. B. C., 1980, The isolation of human erythroyte Band 3 polypeptide labelled with a photosensitive hydrophobic probe, *Biochem. J.* **187**:719–725.

Wieth, J. O., and Bjerrum, P. J., 1982, Titration of transport and modifier sites in the red cell anion transport system, *J. Gen. Physiol.* **79**:253–282.

Wieth, J. O., and Brahm, J., 1983, Cellular anion exchange, in: *Physiology and Pathology of Electrolyte Metabolism* (G. Giebisch and D. Seldin, eds.), Raven Press, New York (in press).

Wieth, J. O., Brahm, J., and Funder, J., 1980, Transport and interactions of anions and protons in the red blood cell membrane, *Ann. N. Y. Acad. Sci.* **341**:394–418.

Wieth, J. O., Bjerrum, P.J., and Borders, C. L., Jr., 1982, Irreversible inactivation of red cell chloride exchange with phenylglyoxal, an arginine-specific reagent, *J. Gen. Physiol.* **79**:283–312.

Williams, D. G., Jenkins, R. E., and Tanner, M. J. A., 1979, Structure of the anion-transport protein of the human erythrocyte membrane. Further studies on the fragments produced by proteolytic digestion, *Biochem. J.* **181**:477–493.

Wolosin, J. M., 1980, A procedure for membrane-protein reconstitution and the functional reconstitution of the anion transport system of the human erythrocyte membrane, *Biochem. J.* **189**:35–44.

Yeltman, D. R., and Harris, B. G., 1980, Localization and membrane association of aldolase in human erythrocytes, *Arch. Biochem. Biophys.* **199:**186–196.

Yu, J., and Steck, T. L., 1975a, Isolation and characterization of Band 3, the predominant polypeptide of the human erythrocyte membrane, *J. Biol. Chem.* **250:**9170–9175.

Yu, J., and Steck, T. L., 1975b, Associations of Band 3, the predominant polypeptide of the human erythrocyte membrane, *J. Biol. Chem.* **250:**9176–9184.

Yu, J., Fischman, D. A., and Steck, T. L., 1973, Selective solubilization of proteins and phospholipids from red blood cell membranes by non-ionic detergents, *J. Supramol. Struct.* **1:**233–248.

Zaki, L., 1981, Inhibition of anion transport across red blood cells with 1,2-cyclohexamedione, *Biochem. Biophys. Res. Commun.* **99:**243–251.

Zaki, L., Fasold, H., Schuhmann, B., and Passow, H., 1975, Chemical modification of membrane proteins in relation to inhibition of anion exchange in human red blood cells, *J. Cell. Physiol.* **36:**471–494.

3

BIOSYNTHESIS AND ASSEMBLY OF MITOCHONDRIAL PROTEINS

Martin Teintze and Walter Neupert

1. INTRODUCTION

Eukaryotic cells are divided into a number of subcellular compartments, each surrounded by one or more membranes. By creating these compartments (i.e., organelles), eukaryotic cells have evolved a host of new capabilities compared to the simpler prokaryotic cells. The creation within the cell of compartments enclosed by membranes allows, for instance, the use of metabolic pathways in opposite directions at the same time, or the storage of substances (such as Ca^{2+} in the sarcoplasmic reticulum) where they can be released when needed to initiate reactions. Proteolytic enzymes can be sequestered where they do not interfere with other cell functions, and ion or proton gradients can be generated across membranes within the cell to drive reactions such as the synthesis of ATP. This increased complexity, however, brings with it a host of new problems for the eukaryotic cell. A major one is that almost all the proteins of the intracellular organelles are synthesized on cytoplasmic ribosomes (Schatz and Mason, 1974; Chua and Schmidt, 1979). Hundreds of different proteins must be directed into their proper organelles during or after translation. The membranes of these organelles, on the other hand, are practically impermeable to macromolecules such as proteins. Therefore mechanisms must exist for specifically transferring newly synthesized proteins into or across the membrane or membranes of the proper organelle.

Most theories that attempt to explain this phenomenon postulate the

Martin Teintze and Walter Neupert • Institute of Biochemistry, University of Göttingen, D-3400 Göttingen, Federal Republic of Germany.

existence of complementary structures on the proteins and the organelles, i.e., receptors in the outer membranes of these organelles. Receptor proteins would be able to bind specifically a protein or class of proteins destined for an organelle during or after its synthesis and mediate its transport into or across the membrane.

Although no such receptor proteins have been isolated yet, the evidence suggesting their existence has been growing (Hennig and Neupert, 1981). The transfer of proteins from their site of synthesis to their functional site is irreversible; this requires that the proteins be synthesized as precursors that are in some way different from the mature functional form of the protein. Some of the ways by which this is accomplished will be discussed in the following sections.

Two kinds of organelles, the mitochondria and the chloroplasts, have their own DNA and ribosomes and are able to synthesize a few of their own proteins. A number of those are subunits of enzyme complexes that also contain subunits coded for on nuclear genes and synthesized on cytoplasmic ribosomes. Very little is known to date about how these two synthetic systems are coordinated so that these enzyme complexes are assembled with stoi-chiometric ratios of subunits.

The study of the molecular mechanisms of intracellular protein transport involves following single, well-defined proteins that can be assigned to one particular organelle over the entire path from the synthetic origin to the functional site. This requires isolation and purification of such proteins and the preparation of specific antibodies with which one can detect the minute quantities of precursors or intermediates that might be involved in the transfer process.

Many studies on intracellular protein transport have focused on mito-chondria. A large number of mitochondrial proteins have been well-charac-terized with respect to structure, localization, and site of synthesis. Mito-chondria can easily be separated from the remaining cell components, allowing separation of the synthesis and transfer processes.

2. POST- VS. COTRANSLATIONAL TRANSPORT

There are probably several different mechanisms by which proteins syn-thesized in the eukaryotic cell reach their final destinations. The proteins whose intracellular transport has been studied can be divided into two major classes: those that are transported cotranslationally and those that are transported posttranslationally.

Cotranslational transport is observed in the endoplasmic reticulum (ER) and probably the plasma membrane of bacteria (Sabatini *et al.*, 1982). Proteins that are destined to be secreted via the Golgi apparatus or have their functional site in the lumen of the endoplasmic reticulum are synthesized by ribosomes bound to the cytoplasmic face of the ER membrane (Palade, 1975; Blobel and Dobberstein, 1975). The nascent polypeptide chain is discharged through the membrane during protein synthesis, before the polypeptide is fully folded.

Some details of the cotranslational mechanism have recently been elucidated. An 11 S protein composed of six polypeptides was purified from dog pancreas rough microsomes and was shown to bind to polysomes synthesizing secretory proteins. Since it apparently recognizes the amino terminal signal sequence of the nascent polypeptide, it was termed the "signal recognition protein" (SRP) (Walter *et al.*, 1981). This protein complex apparently also mediates the binding of the polysomes synthesizing the secretory protein to microsomal membranes (Walter and Blobel, 1981a) and stops chain elongation until the SRP–ribosome complex binds to the microsomal membrane (Walter and Blobel, 1981b). The receptor for the SRP or "docking protein" has been identified as an integral glycoprotein (M_r 72,000) of the ER membrane (Walter *et al.*, 1979; Meyer and Dobberstein, 1980; Meyer *et al.*, 1982). Furthermore, the rough ER contains ribosome binding sites called ribophorins (Kreibich *et al.*, 1978) that may serve to keep the ribosome attached to the membrane while elongation of the polypeptide is completed. During this second elongation phase the polypeptide traverses the membrane and a "signal peptidase" removes the signal sequence from the nascent chain (Jackson and Blobel, 1977). A "leader peptidase" has also been isolated from *E. coli* inner and outer membranes (Zwizinski and Wickner, 1980).

On the other hand, the intracellular transport of a large number of proteins into mitochondria, as well as chloroplasts, glyoxysomes, and peroxisomes, has been shown to take place by a posttranslational mechanism (Chua and Schmidt, 1979; Neupert and Schatz, 1981). This means that synthesis of the precursor protein is separated in both space and time from the transfer into the organelle rather than being obligatorily coupled as in cotranslational transport.

Two different types of experiments have been employed to follow the biogenesis of a mitochondrial protein. In *in vivo* pulse-labeling studies one can often detect the precursors, if they differ in molecular weight or immunological properties from the mature proteins, and follow their processing by lysing the cells at different time points and immunoprecipitating the proteins. These experiments show that processing and transfer still occur when protein synthesis has been stopped with cycloheximide, and that labeled precursors appear in the cytosol before the labeled proteins appear in the mitochondrial fraction (Hallermayer et al., 1977; Schatz, 1979; Teintze et al., 1982). It is difficult to tell on what type of ribosome a protein is made, however, or where it is localized in the cell.

Experiments can also be done in vitro using precursors synthesized in a homologous or heterologous cell-free system and isolated mitochondria. These can be resuspended in the postribosomal supernatant of the cell-free system after protein synthesis has been completed, and import of precursor proteins into the mitochondria in the absence of protein synthesis can be observed (Korb and Neupert, 1978; Maccecchini et al., 1979a; Zimmermann and Neupert, 1980). In this type of experiment one can distinguish proteins that remain in solution, those that are bound to recognition sites on the mitochondria but still accessible to externally added proteases, and those that have been imported into a protease-resistant and/or functional location in the mitochondria. The results show that as the precursors disappear from the supernatant they are first bound to the outside of the mitochondria and then transferred to their functional site. It is during this second step that they are converted to their mature, functional form.

So far almost all mitochondrial precursor proteins studied are synthesized on free polysomes and are transferred posttranslationally (Table I). However, general rules for transfer into organelles cannot be drawn. In fact, insertion of proteins into ER or plasma membranes appears to obey both mechanisms, depending on the protein (Sabatini et al., 1982). Transfer of proteins into the plasma membrane of bacteria may also occur by both cotranslational and posttranslational pathways (Wickner, 1980).

3. PRECURSOR FORMS OF MITOCHONDRIAL PROTEINS

The mitochondrial proteins that are made on cytoplasmic ribosomes are synthesized as precursors that can be immunoprecipitated after pulse-labeling cells in vivo or after synthesis in an in vitro cell-free system. Many of these

TABLE I. Mitochondrial Precursor Proteins

Protein	Organism or tissue	Location in mitochondrion	Site of synthesis (polysomes)	Posttranslational transfer shown	Energy-dependent	Precursor MW (approx.)	Mature MW (approx.)	Reference(s)
Porin	Neurospora	Outer membrane	Free	Yes	No	31,000	31,000	Freitag et al. (1982a,b)
Cytochrome b_2	Yeast	Intermembrane space	N.D.[a]	Yes	Yes	68,000	58,000	Gasser et al. (1982)
Cytochrome c peroxidase	Yeast	Intermembrane space	N.D.	Yes	No	39,500	33,500	Maccecchini et al. (1979b), Nelson and Schatz (1979)
Sulfite oxidase	Rat liver	Intermembrane space	Free	No	N.D.	59,000	55,000	Mihara et al. (1982)
Cytochrome c	Neurospora	Intermembrane space–inner membrane	Free	Yes	No	12,000	12,000	Korb and Neupert (1978), Zimmermann et al. (1979a, 1981)
Cytochrome c	Rat liver	Intermembrane space–inner membrane	Free	Yes	No	12,000	12,000	Matsuura et al. (1981)
ADP/ATP carrier	Neurospora	Inner membrane	Free	Yes	Yes	32,000	32,000	Zimmermann et al. (1979b), Zimmermann and Neupert (1980)
ATPase subunit 9	Neurospora	Inner membrane	Free	Yes	Yes	12,000	8,000	Michel et al. (1979), Zimmerman et al. (1981)
Cytochrome bc_1 complex	Neurospora	Inner membrane	N.D.	Yes	Yes	51,500	50,000	Teintze et al. (1982)
Subunit I			N.D.	Yes	Yes	47,500	45,000	
Subunit II			Free	Yes	Yes	38,000	31,000	
Cytochrome c_1			N.D.	Yes	Yes	28,000	25,000	
Subunit V			N.D.	No	N.D.	14,000	14,000	
Subunit VI			N.D.	Yes	Yes	12,000	11,500	
Subunit VII			N.D.	No	N.D.	11,600	11,200	
Subunit VIII								
Cytochrome bc_1 complex	Yeast							Côté et al. (1979), Nelson and Schatz (1979)
Cytochrome c_1			N.D.	No	Yes	37,000	31,000	
Subunit V			N.D.	No	Yes	27,000	25,000	

(*Continued*)

TABLE I. *(Continued)*

Protein	Organism or tissue	Location in mitochondrion	Site of synthesis (polysomes)	Posttranslational transfer shown	Energy-dependent	Precursor MW (approx.)	Mature MW (approx.)	Reference(s)
Cytochrome c oxidase	Rat liver	Inner membrane						Schmelzer and Heinrich (1980), Heinrich (1982)
Subunit IV			Free + bound	No	N.D.	19,500	16,500	
Subunit V			Free + bound	No	N.D.	15,500	12,500	
Cytochrome c oxidase	Yeast	Inner membrane						Lewin et al. (1980), Mihara and Blobel, (1980)
Subunit IV			N.D.	No	N.D.	—	—	
Subunit V			N.D.	Yes	N.D.	—	—	
Subunit VI			N.D.	No	N.D.	—	—	
Subunit VII			N.D.	No	N.D.	20,000	12,500	
Cytochrome P450$_{scc}$	Bovine adrenal cortex	Inner membrane	N.D.	No	N.D.	54,500	49,000	DuBois et al. (1981)
Cytochrome P450$_{11-\beta}$	Bovine adrenal cortex	Inner membrane	Free + bound	No	N.D.	50,000	45,000	Nabi et al. (1980)
F$_1$-ATPase	Yeast	Matrix–inner membrane						Maccecchini et al. (1979a), Nelson and Schatz (1979), Lewin et al. (1980)
α Subunit			N.D.	Yes	Yes	64,000	58,000	
β Subunit			N.D.	Yes	Yes	56,000	54,000	
γ Subunit			N.D.	Yes	Yes	40,000	34,000	
Adrenodoxin	Bovine adrenal cortex	Matrix	Free + bound	Yes	N.D.	20,000	12,000	Nabi and Omura (1980)
δ-Aminolevulinic acid synthetase	Rat liver	Matrix	Free	No	N.D.	51,000	45,000	Yamauchi et al. (1980a,b)
Aspartate aminotransferase	Chicken heart	Matrix	Free	No	N.D.	47,500	44,500	Sonderegger et al. (1980, 1982)
Carbamoyl-phosphate synthetase	Rat liver	Matrix	Free	Yes	N.D.	47,000	45,000	Sakakibara et al. (1980), Mori et al. (1979), Raymond and Shore (1979, 1980), Shore et al. (1979)
	Rat liver	Matrix	Free	Yes	N.D.	165,000	160,000	

Citrate synthase	Neurospora	Matrix	N.D.	No	47,000	44,500	Harmey and Neupert (1979)
L-Glutamate dehydrogenase	Rat liver	Matrix	Free	No	60,000	54,000	Mihara et al. (1982)
D-β-Hydroxybutyrate dehydrogenase	Rat liver	Matrix	Free	No	37,000	32,000	Mihara et al. (1982)
Malate dehydrogenase	Rat liver	Matrix	Free	No	38,000	37,000	Mihara et al. (1982)
Ornithine carbamoyltransferase	Rat liver	Matrix	N.D.	Yes	39,400 / 43,000	36,000 / 39,000	Mori et al. (1980), Conboy and Rosenberg (1981), Morita et al. (1982)
Superoxide dismutase	Yeast	Matrix	N.D.	No	26,000	24,000	Autor (1982)

[a] N.D., not determined

precursors have an additional sequence that is removed by a proteolytic enzyme during or after the transfer process. Their apparent molecular weights are larger than those of the mature proteins by anywhere from 500 daltons for subunit VII of the *Neurospora* cytochrome bc_1 complex (Teintze *et al.*, 1982) to 10,000 daltons for cytochrome b_2 (Gasser *et al.*, 1982). Other proteins for which this type of transfer with proteolytic processing has been observed are listed in Table I.

For cytochrome c_1 in *Neurospora* (Teintze *et al.*, 1982) and for cytochromes c_1 and b_2 in yeast (Gasser *et al.*, 1982), the proteolytic processing takes place in two separate steps via a form with intermediate molecular weight.

Some mitochondrial proteins, however, are synthesized with the same molecular weight as the mature functional protein and are transferred without proteolytic processing. These include cytochrome c (Zimmermann *et al.*, 1979a), the ADP/ATP carrier (Zimmermann *et al.*, 1979b), subunit VI of the cytochrome bc_1 complex (Teintze *et al.*, 1982), and the outer membrane porin (Freitag *et al.*, 1982b). In these cases there must be some kind of conformation change that allows the proteins to get into or through the membrane and prevents them from leaving their compartment.

The most extensively studied example of this type of transfer process is that of cytochrome c, because it is the only imported protein whose precursor, apocytochrome c, has been prepared in large quantities. Apocytochrome c is transferred across the outer membrane into the intermembrane space in conjunction with the covalent attachment of a heme group to form mature holocytochrome c (Hennig and Neupert, 1981). The heme group can be removed from isolated holocytochrome c to produce chemical quantities of apocytochrome c, which is identical to the precursor synthesized in the cell. Antibodies prepared against *Neurospora* apo- and holocytochrome c do not cross-react (Korb and Neupert, 1978). Excess unlabeled apocytochrome c, but not holocytochrome c, can compete with radioactive apocytochrome c for binding and transfer into *Neurospora* mitochondria (Hennig and Neupert, 1981).

Thus, upon addition of the heme group, the protein probably undergoes a conformational change that results in transfer of the holocytochrome c into the intermembrane space.

No covalent modifications have been detected so far for any of the other mitochondrial protein precursors that are imported without proteolytic processing. The precursor of the ADP/ATP carrier of *Neurospora* mitochondria synthesized *in vitro* in a heterologous cell-free system has the same apparent molecular weight on sodium dodecylsulfate gels as the mature protein and was shown not to have an additional sequence at the N terminal by translation in the presence of N-formyl[^{35}S]methionyl-tRNA (Zimmermann *et al.*, 1979b). The precursors were found to be present as soluble aggregates with molecular weights in the range of 100,000 to 500,000 (Zimmermann and Neupert, 1980), whereas the mature protein is an insoluble integral inner membrane protein with a molecular weight of about 30,000.

The outer membrane porin of *Neurospora* mitochondria, which is also synthesized with the same molecular weight as the mature protein (Freitag *et al.*, 1982b), has recently been isolated in a water-soluble lipid-free form (H. Freitag and W. Neupert, unpublished observations). This water-soluble form of porin can compete with the *in-vitro*-synthesized precursor for binding and transfer into mitochondria, unlike isolated porin, which contains lipid and can only be solubilized in the presence of detergent.

4. EVIDENCE FOR THE EXISTENCE OF SPECIFIC RECEPTORS

The precursor proteins that are synthesized on free cytoplasmic ribosomes must have a method of recognizing the organelle for which they are intended and a method of entering or crossing a membrane that is normally impermeable to proteins. The most logical mechanism for such a process is to have specific receptors in the outer membrane of the organelle that can both bind the precursor and mediate its transfer.

The evidence for the presence of a specific receptor in the outer mitochondrial membrane is strongest in the case of cytochrome c. The presence of specific saturable binding sites is indicated by the following data. Excess unlabeled apocytochrome c can compete with labeled apocytochrome c and inhibit its transfer into the mitochondria (Hennig and Neupert, 1981). Deuterohemin, an analog of the natural prosthetic group protohemin lacking the vinyl groups necessary for covalent attachment to the apoprotein, inhibits transfer but not binding. In the presence of deuterohemin, radioactive apocytochrome c bound to *Neurospora* mitochondria can be displaced by excess unlabeled apocytochrome c from *Neurospora* (Hennig and Neupert, 1981). Apocytochrome cs from other species compete less effectively in a manner that parallels their sequence homology with the native *Neurospora* protein, with an apocytochrome c from a bacterium *(Paracoccus denitrificans)* not able to compete for the binding sites at all (B. Hennig and W. Neupert, unpublished observations). If mitochondria with radiolabeled apocytochrome c bound in the presence of deuterohemin are isolated, washed, and placed in a medium containing protohemin, the natural prosthetic group, the apocytochrome c is converted to holocytochrome c and transferred into the mitochondria (Hennig and Neupert, 1981). This shows that the apocytochrome c is initially bound to a location on the outside of the mitochondria from which it can be displaced by externally added unlabeled apocytochrome c, yet it is bound to a location from which it can be transferred into the mitochondria when the inhibition by deuterohemin is relieved.

Detailed binding studies have been carried out using [^{14}C]apocytochrome c prepared by reductive methylation (H. Köhler, B. Hennig and W. Neupert, unpublished observations). The binding of labeled apocytochrome c to whole mitochondria gives a biphasic Scatchard plot with about 90 pmoles of high-affinity binding sites ($K_d = 10^{-7}$ M) per mg of mitochondrial protein. This is a very large number considering the low protein content of the outer mitochondrial membrane. In fact, the only outer membrane protein present in sufficient quantity to explain the number of binding sites found is the porin (Freitag et al., 1982a). Studies are now under way to determine whether the porin could be the receptor for apocytochrome c.

The outer membrane of mitochondria also does not appear to have enough proteins for there to be a separate receptor for each of the many different cytoplasmically synthesized proteins that must be bound and imported. However, efforts to find another protein that uses the same receptor as cytochrome c have proved fruitless. In experiments using Neurospora mitochondria and radioactive precursor proteins synthesized in reticulocyte lysates, excess unlabeled apocytochrome c, at a concentration sufficient to inhibit the import of radioactive cytochrome c, had no effect on the transfer of the ADP/ATP carrier, subunit 9 of the oligomycin-sensitive ATPase, cytochrome c_1, subunit V of the cytochrome bc_1 complex, or the outer membrane porin (Zimmermann et al., 1981; Teintze et al., 1982; M. Teintze, H. Freitag, and W. Neupert, unpublished observations).

Although studies of binding to putative receptors by precursors of mitochondrial proteins other than cytochrome c are hampered by the fact that the precursors cannot be obtained in sufficient quantity to saturate the binding, there is nevertheless some evidence for the existence of receptors that may be involved in the transfer of the ADP/ATP carrier and that of subunit 9 of the oligomycin-sensitive ATPase. Treatment of Neurospora mitochondria with trypsin prior to incubation with the reticulocyte lysate containing the precursors blocks the import of ATPase subunit 9, indicating that a protein in the outer membrane is probably involved in its transfer (B. Schmidt and W. Neupert, unpublished observations). Also, the precursors to both proteins can be bound to mitochondria in the presence of antimycin A and oligomycin, which break down the membrane potential (see Section 5) and thereby prevent transfer. The bound precursors can be degraded by adding a protease such as proteinase K. However, if the mitochondria with the bound precursors are reisolated, washed, and resuspended in a medium containing N,N,N',N'-tetramethylphenylenediamine (TMPD) and ascorbic acid, which reenergize the mitochondria, the precursors are transferred to a protease-resistant location in the mitochondria and converted to their mature forms (as determined by carboxyatractyloside (CAT) binding ability for the ADP/ATP carrier and by processing to the size of the mature protein for ATPase subunit 9) (M. Schleyer, B. Schmidt, and W. Neupert, unpublished observations). These results indicate that the precursors were bound to a location on the outside of the mitochondria, probably to a protein, from which they could then be imported when the membrane potential was restored.

The precursor of the outer membrane porin from *Neurospora* synthesized in a reticulocyte lysate in the presence of [^{35}S]methionine binds to mitochondria rapidly at 4°C but is transferred only very slowly. If the mitochondria are reisolated after a short incubation with precursor at 4°C and resuspended in unlabeled reticulocyte lysate at 25°C, the radioactive porin is inserted into the outer membrane and becomes protease-resistant like the mature protein (H. Freitag and W. Neupert, unpublished observations).

5. TRANSFER OF MANY PROTEINS REQUIRES A MEMBRANE POTENTIAL

The posttranslational transfer of many cytoplasmically synthesized proteins into mitochondria is energy dependent. This has been shown for the ADP/ATP carrier and subunit 9 of the oligomycin-sensitive ATPase; cytochrome b_2; cytochrome c_1; subunits I, II, V, and VII of the cytochrome bc_1 complex; and ornithine carbamoyltransferase in *in vitro* transfer experiments (Table I). In addition, the transfer of subunits α, β, and γ of F_1-ATPase, ATPase subunit 9, cytochrome b_2, cytochrome c_1, subunit V of the cytochrome

bc_1 complex, and superoxide dismutase has been shown to be energy dependent using *in vivo* pulse-labeling experiments (Table I). In most of these experiments, uncouplers or ionophores were used to break down the potential across the mitochondrial inner membrane and this resulted in an inhibition of protein transport and proteolytic processing. The absence of a membrane potential, however, also leads to a depletion of the intramitochondrial ATP because the oligomycin-sensitive ATPase tries to compensate by pumping out protons (Stigall *et al.*, 1979; Scarpa, 1979). This ATPase activity is strongly inhibited by oligomycin (Schleyer *et al.*, 1982). In order to determine whether it is the membrane potential or ATP that is required for the import of these proteins into mitochondria, the transfer has been studied *in vitro* under conditions in which either the membrane potential is broken down but ATP is still present or the membrane potential is maintained but the intramitochondrial ATP is depleted. The transfer of the ADP/ATP carrier and ATPase subunit 9 was studied in a system employing isolated *Neurospora* mitochondria suspended in the postribosomal supernatant of a reticulocyte lysate that had been incubated with *Neurospora* RNA and [^{35}S]methionine (Schleyer *et al.*, 1982). The uncouplers and protonophores carbonylcyanide *m*-chlorophenylhydrazone (CCCP) and dinitrophenol were used. Both with and without oligomycin and ATP the transfer of the precursor proteins and the processing of ATPase subunit 9 were inhibited, but binding of the precursors to the mitochondria still occurred. Valinomycin, a K^+ ionophore that breaks down the membrane potential in the presence of high K^+ concentrations like that present in the reticulocyte lysate (90 mM), gave the same results as CCCP. In both of these experiments the intramitochondrial ATP level was high, as demonstrated by the protein synthesis on the mitochondrial ribosomes (Schleyer *et al.*, 1982). This is due to three factors: (1) the inhibition of ATPase by oligomycin, (2) the high ATP concentration in the lysate, and (3) the absence of a membrane potential, which causes the external and internal ATP pools to equilibrate via the ADP/ATP carrier (Heldt *et al.*, 1972). In another experiment carboxyatractyloside, which blocks the ADP/ATP carrier, was added together with oligomycin to deplete the intramitochondrial ATP pool while leaving the membrane potential intact (Schleyer *et al.*, 1982). Under these conditions there was no inhibition of transfer or proteolytic processing. The same experiments have since been carried out on cytochrome c_1 and subunits I, V, and VII of the cytochrome bc_1 complex (Teintze *et al.*, 1982). In each case the same pattern of response to inhibitors was observed. These observations show that the membrane potential, and not ATP, is the source of energy for the import of precursor proteins into mitochondria. The ionophore nigericin, which exchanges K^+ for H^+ and causes a breakdown of the proton gradient without affecting the membrane potential (Graven et al., 1966; Pressman, 1976), did not inhibit the transfer of the ADP/ATP carrier or the transfer and processing of ATPase subunit 9 (Schleyer et al., 1982). Apparently it is the electrical membrane potential and not just the proton gradient that is required.

The energy-dependent transfer of precursor proteins can also be blocked by a combination of antimycin A, an inhibitor of the electron transport chain, and oligomycin to inhibit the oligomycin-sensitive ATPase, thus blocking both pathways for generating a membrane potential. This inhibition can be relieved by the addition of ascorbate and TMPD, which feed electrons directly into the cytochrome oxidase, i.e., after the site of inhibition of antimycin A, and allow the membrane potential to be reestablished. If precursors are bound in the presence of antimycin A and oligomycin and the mitochondria are reisolated and resuspended in a medium containing TMPD and ascorbate but no additional precursor, the bound precursors are imported and converted to the mature proteins (Schleyer et al., 1982).

Schatz and co-workers studied the transfer of subunits α, β, and γ of F_1-ATPase, cytochrome c_1, and subunit V of the cytochrome bc_1 complex in vivo and rho⁻ mutants of S. cerevisiae and came to the conclusion that ATP was required rather than the membrane potential (Nelson and Schatz, 1979). Rho⁻ mutants have mitochondria deficient in protein synthesis and thus lack both a functional respiratory chain and functional oligomycin-sensitive ATPase (Lloyd, 1974). These mitochondria should therefore not be able to generate a membrane potential by respiration or ATP hydrolysis. Nevertheless, rho⁻ cells contain mitochondria that import and process cytoplasmically synthesized proteins. Thus it was concluded that ATP obtained from the cytoplasm via the ADP/ATP carrier (Subik et al., 1974) must be providing the necessary energy (Nelson and Schatz, 1979). Later studies on in vitro import of cytochrome b_2 into isolated yeast mitochondria, however, suggested that this protein requires an electrochemical gradient for import (Gasser et al., 1982). The situation in the rho⁻ mutants remains unclear; however, one possibility is that the import of ATP coupled to the export of ADP generates a small membrane potential in the mitochondria because it is an electrogenic process (Scarpa, 1979).

Both cytochrome c and the porin of the outer mitochondrial membrane do not require energy for their import (Zimmermann et al., 1981; Freitag et al., 1982b). Both are not proteolytically processed, but neither is the ADP/

ATP carrier, which nevertheless requires energy for import. The reason for the lack of an energy requirement for import of cytochrome c, which is located in the intermembrane space, and the outer membrane porin may be the fact that neither needs to be transferred into or across the inner mitochondrial membrane. Cytochrome b_2 in yeast, an intermembrane space enzyme that does require an inner membrane potential for import and processing, is thought to cross and recross the inner mitochondrial membrane before reaching its functional site. The same may be true for cytochrome c_1, which is located with its major domain on the cytoplasmic side of the inner membrane (Gasser et al., 1982).

6. PROTEOLYTIC PROCESSING ENZYMES

There have been a number of reports recently describing proteolytic enzymes isolated from mitochondria that process precursors of mitochondrial proteins to the apparent molecular weights of the mature proteins. Böhni et al. (1980) discovered a protease activity in hypotonic extracts of yeast and rat liver mitochondria that apparently correctly processed the precursor of yeast F_1-ATPase subunits α, β, and γ and cytochrome c oxidase subunit V synthesized in an in vitro cell-free system. All four of these proteins are synthesized as larger precursors on cytoplasmic ribosomes and then posttranslationally transferred to their functional sites in the mitochondrial matrix or inner membrane (Maccecchini et al., 1979a; Lewin et al., 1980). Both yeast and rat liver processing enzyme activities could be inhibited by the metal chelators o-phenanthroline and ethylenediaminetetraacetic acid (EDTA) but not by serine protease inhibitors; the processing activity of the rat liver mitochondria fractionated with marker enzymes of the matrix space (Böhni et al., 1980). The yeast mitochondrial protease apparently also processes the precursors of cytochrome c_1 and cytochrome b_2, an intermembrane space enzyme, but only to their intermediate forms (Gasser et al., 1982). The proteolytic processing enzyme of the yeast mitochondria was further purified and characterized by McAda and Douglas (1982). They showed that the enzyme that processes subunit 2 (β) of the F_1-ATPase in yeast was a metallo-endoprotease complex of M_r 115,000 with a pH optimum between 7 and 8. Its activity was inhibited by EDTA and o-phenanthroline, and could be restored with excess Co^{2+} or Mn^{2+}.

In studies on rat liver mitochondria, two different groups have isolated a protease from the mitochondrial matrix that processes the precursor of ornithine carbamoyltransferase (Miura *et al.*, 1982; Conboy *et al.*, 1982). The protease characterized by Miura *et al.* (1982) had an M_r of 108,000 and was inhibited by the metal chelators EDTA, *o*-phenanthroline, and zincon. It processed the precursor of ornithine carbamoyltransferase (M_r 39,400) to an intermediate form (M_r 37,000) but not to the size of the mature subunit (M_r 36,000). This processing enzyme was also isolated from the mitochondrial matrix of rat kidney, spleen, heart, and ascites tumor cells, all of which lack ornithine carbamoyltransferase, and therefore probably has a broader specificity. The enzyme isolated by Conboy *et al.* (1982) converts pre-ornithine carbamoyltransferase to the apparent molecular weight of the mature subunit as well as to the intermediate form. It was shown to require Zn^{2+}, Co^{2+}, or Mn^{2+} for activity but still processed the precursor of ornithine carbamoyltransferase to the intermediate form in the absence of these metal ions.

Considering the common characteristics of all the precursor-processing enzymes described above, it would be interesting to know whether there is only one such enzyme, whether this enzyme can process precursors other than those described above, and whether the same enzyme can be found in the mitochondria of other species. Sequencing analyses will be required, however, to show that any such protease actually processes the precursors correctly.

7. DIFFERENT ORGANISMS HAVE CLOSELY RELATED TRANSFER MACHINERIES

Labeled precursors to *Neurospora* mitochondrial proteins synthesized in a reticulocyte lysate cell-free system can be imported into isolated mitochondria from other organisms and are converted to their mature forms. The transfer of the *Neurospora* ADP/ATP carrier and ATPase subunit 9 into rat liver mitochondria requires an electrochemical gradient across the inner mitochondrial membrane just like the corresponding transfer into *Neurospora* mitochondria (Schleyer *et al.*, 1982). Both proteins were transferred into protease-resistant locations and the ATPase subunit 9 was processed to the molecular weight of the mature subunit. Similar results have been obtained with the transfer of these proteins into yeast and guinea pig heart mitochondria (B. Schmidt, M. Schleyer, and W. Neupert, unpublished observations). The *Neu-*

rospora ADP/ATP carrier was shown to be in its functional form after transfer into yeast mitochondria on the basis of its ability to bind CAT, which the precursor does not. The ATPase subunit 9 precursor was processed correctly by the yeast mitochondria during transfer, as shown by radiosequencing analysis (B. Schmidt, B. Hennig, and W. Neupert, unpublished observations). This is especially interesting since the yeast ATPase subunit 9 is encoded on the mitochondrial genome and synthesized inside the mitochondria without an additional sequence (Sebald and Wachter, 1978; Macino and Tzagoloff, 1979; Hensgens *et al.*, 1979). Yeast mitochondria therefore do not have to import and process a cytoplasmically made ATPase subunit 9 precursor *in vivo*, but the mechanism for doing so must be conserved. The precursor of subunit V (Fe/S protein) of the *Neurospora* cytochrome bc_1 complex is also imported into yeast mitochondria and processed to the molecular weight of the mature protein when an inner membrane potential is present (Teintze *et al.*, 1982). Recently the *Neurospora* outer membrane porin precursor was also successfully imported into rat liver and yeast mitochondria (H. Freitag and W. Neupert, unpublished observations).

The precursor of rat liver ornithine carbamoyltransferase is also imported and processed to the correct molecular weight by mouse liver and kidney mitochondria (Morita *et al.*, 1982).

Combined with the data discussed in the previous section showing that a protease isolated from rat liver processes yeast precursors (Böhni *et al.*, 1980), these results suggest that the mechanism for importing cytoplasmically synthesized precursors into mitochondria including the proteins mediating recognition, import, and processing has been highly conserved in evolution.

8. FUNCTIONAL ASSEMBLY OF MITOCHONDRIAL ENZYME COMPLEXES

Only a very small number of mitochondrial proteins are encoded on the mitochondrial genome (Tzagoloff *et al.*, 1979). Most of these proteins are subunits of enzyme complexes that also contain subunits synthesized in the cytoplasm, such as the oligomycin-sensitive ATPase, cytochrome *c* oxidase, and the cytochrome bc_1 complex. The question of how the synthesis of subunits in the mitochondrial and the cytoplasm is regulated so that stoichiometric amounts of each subunit are available for assembly into a functional complex remains unanswered. It is known only that inhibition of mitochondrial protein synthesis has no gross effect on the production of cytoplasmically synthesized subunits. For instance, *Neurospora* cells can be grown for many generations in the presence of chloramphenicol, which inhibits mitochondrial but not cytoplasmic protein synthesis. The cells still make mitochondria, and even though they contain very little cytochrome *b* (a mitochondrial gene product), cytochrome c_1 is still synthesized in the cytoplasm, imported into the mito-chondria in normal quantities, and proteolytically processed to the mature form, although it cannot be assembled into a cytochrome bc_1 complex (Weiss and Kolb, 1979).

Another problem that has not been completely solved is to demonstrate that precursors of nuclear-coded subunits synthesized in a cell-free system and then transferred into isolated mitochondria are actually assembled into functional complexes. Following transfer *in vitro* of radiolabeled *Neurospora* precursor into mitochondria, the labeled subunit 9 of the oligomycin-sensitive ATPase can be precipitated with antibodies directed against the F_1 portion of the complex (B. Schmidt and W. Neupert, unpublished observations). These antibodies will not precipitate isolated ATPase subunit 9, so the imported subunits must be assembled with the F_1 subunits, but one cannot rule out a subunit exchange during the immunoprecipitation process.

After transfer *in vitro*, the precursor to the ADP/ATP carrier acquires certain characteristics of the dimeric functional carrier (M. Schleyer and W. Neupert, unpublished observations). It binds CAT, as demonstrated by pro-tection against protease activity; CAT is bound by the dimeric carrier in the membrane (Klingenberg *et al.*, 1979) but not by the precursor in the reti-culocyte lysate, or the precursor bound to mitochondria in the presence of energy inhibitors. Also, both the mature assembled ADP/ATP carrier and the protein transferred *in vitro* pass through hydroxyapatite columns, whereas the precursor binds.

Assembly *in vitro* has been clearly demonstrated for ribulose-1,5-bis-phosphate carboxylase in chloroplasts (Chua and Schmidt, 1978). The small subunit of this enzyme is made on free cytoplasmic ribosomes and transferred into the chloroplast, where it is proteolytically processed and assembled with the large subunit, which is made in the chloroplast. After synthesis and transfer *in vitro* it was shown that 80% of the labeled small subunit had been assembled with the large subunit to form the holoenzyme, as determined by density gradient centrifugation and electrophoresis on nondenaturing gels.

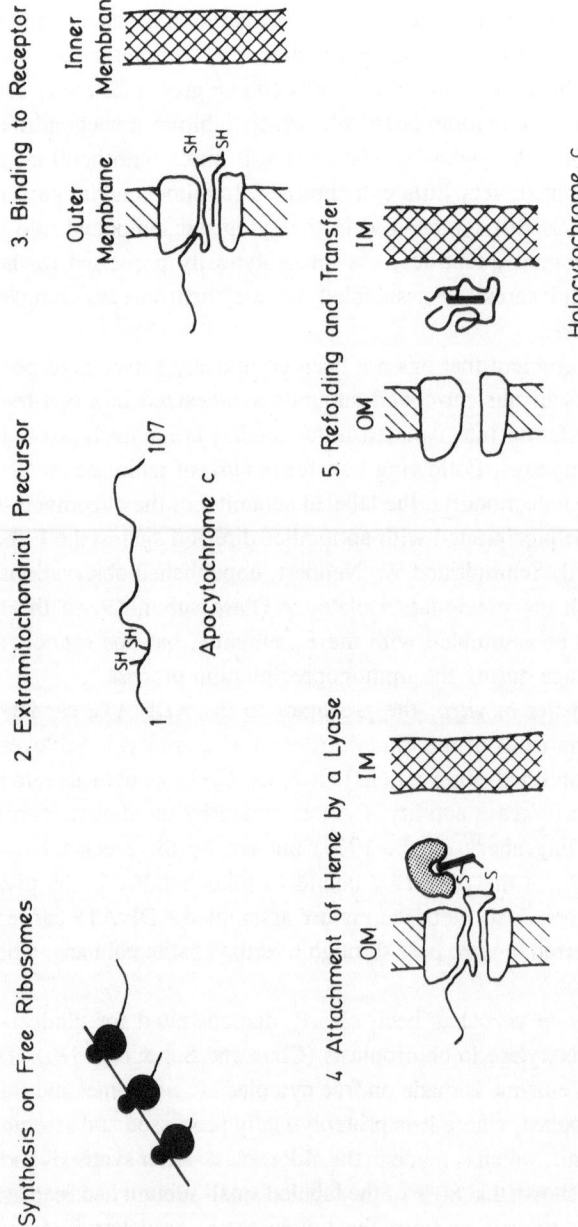

FIGURE 1. Proposed assembly mechanism for cytochrome c.

9. POSSIBLE ASSEMBLY PATHWAYS

The mitochondrial protein whose biogenesis is best understood is cytochrome c. This enzyme is synthesized as apocytochrome c in the cytoplasm and must cross the outer mitochondrial membrane and gain a covalently attached heme group to reach its functional location in the intermembrane space as holocytochrome c bound to the cytoplasmic face of the inner membrane. A schematic representation of the mechanism by which this is thought to occur is shown in Figure 1. Apocytochrome c is synthesized on free cytoplasmic ribosomes with the same amino acid sequence as the mature protein (Zimmermann *et al.*, 1979a). It then binds to specific sites on the outside of the mitochondria before being transferred across the outer membrane in conjunction with the covalent attachment of a heme group via two cysteine residues (Hennig and Neupert, 1981). When labeled apocytochrome c is bound to its receptor, it is still accessible to the outside since it can be displaced specifically by unlabeled apocytochrome c. It must also be accessible to heme and the enzyme that attaches it to the protein. Probably the portion of the apocytochrome c to which the heme must be attached extends into the intermembrane space so that the change in conformation induced by attachment of the heme group pulls the rest of the protein into the intermembrane space. This mechanism does not require the presence of any type of "signal" sequence in the protein; the receptor recognizes the conformation of apocytochrome c but not holocytochrome c, and the change in conformation upon attachment of the heme group is probably responsible for both the transfer across the membrane and for the irreversibility of the import process. The portion of apocytochrome c responsible for recognition may be located in the highly conserved sequence around residues 70–80, which is rather hydrophobic. This was concluded from competition experiments with fragments of apocytochrome c (Matsuura *et al.*, 1981; B. Hennig and W. Neupert, unpublished observations).

The simplest mechanism in mitochondrial membrane assembly should be the insertion of cytoplasmically synthesized proteins into the outer membrane, since such proteins do not have to cross one or more membranes to reach the functional site. The only well-characterized protein of the outer mitochondrial membrane whose biogenesis has been studied is the porin. The outer membrane porin is synthesized on free cytoplasmic ribosomes as a water-soluble precursor that has the same molecular weight as the mature protein

(Freitag *et al.*, 1982b). Also, in this case, specific binding of the precursor precedes its insertion into the membrane. Isolated detergent-solubilized mature porin will insert into artificial lipid bilayers to form pores (Freitag *et al.*, 1982a). The water-soluble precursor form can be regenerated from the isolated porin by removing all the detergent. This lipid-free form can compete with *in-vitro*-synthesized precursor for binding and transfer into the mitochondrial outer membrane, but it will not insert into artificial lipid bilayers until sterols are added (H. Freitag, R. Benz, and W. Neupert, unpublished observations). This suggests that the composition of the membrane may be important for the recognition or insertion process and that no covalent modification is necessary for the assembly of the porin.

The insertion of proteins into the outer mitochondrial membrane is topologically similar to the insertion of proteins into the plasma membrane or microsomal membranes. Posttranslational transfer into these compartments has also been observed. Cytochrome b_5 and cytochrome b_5 reductase are synthesized on free polysomes without any additional sequences (Okada *et al.*, 1982) and the large unglycosylated subunit of Na^+, K^+-ATPase is also made on free polysomes before assembly into the plasma membrane, although the small glycosylated subunit is synthesized in the rough endoplasmic reticulum (Sabatini *et al.*, 1982).

The transfer of proteins from the cytoplasm to the inner mitochondrial membrane or the matrix space is more complicated, since it involves crossing the outer membrane, the aqueous intermembrane space, and then the inner membrane. This process may occur via some type of contact site between the inner and outer membranes (Figure 2). Almost all precursors of inner membrane and matrix proteins whose sites of synthesis have been investigated are made on free polysomes (Table I). In most cases a cytoplasmic precursor pool has been observed and the import into mitochondria is posttranslational. The transfer of proteins into the inner membrane or the matrix requires energization of the mitochondria (Table I) and for a number of inner membrane proteins it has been shown that it is the electrochemical gradient across the inner membrane that drives the import and processing (Schleyer *et al.*, 1982; Teintze *et al.*, 1982). Most inner membrane and matrix proteins are made as larger precursors, whose additional sequences are proteolytically cleaved sometime during the transfer and assembly process. A metal-dependent endoprotease that processes some of these precursors to their mature size has been isolated from the mitochondrial matrix by a number of groups. As might

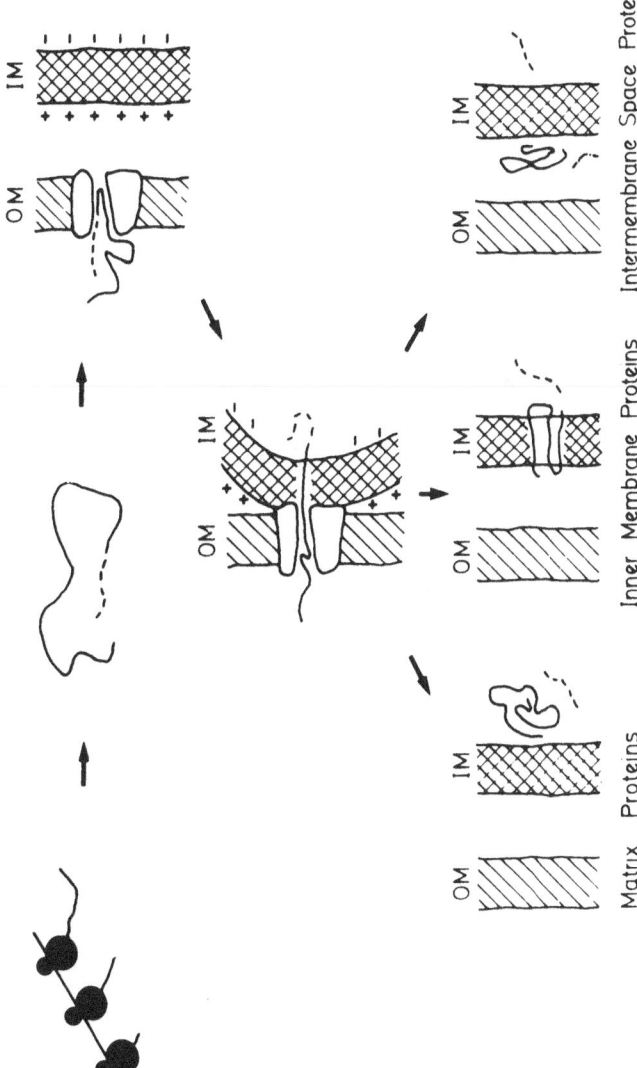

FIGURE 2. Hypothetical pathway for assembly of mitochondrial proteins requiring energy for transfer. The dashed and dotted lines represent portions of the sequence that are removed during transfer. OM, outer membrane; IM, inner membrane.

be expected, the proteins correctly processed by this enzyme are located in the matrix or on the matrix side of the inner membrane.

The function of the additional sequences on the precursors is not altogether clear. They probably serve to keep the precursors in a different (i.e., water soluble) conformation outside the mitochondria, and removal of the extra sequences inside the mitochondria probably makes the assembly process irreversible. The additional sequence may also be important for the recognition of the precursor by a receptor on the mitochondrial surface.

Cytochrome c_1 (located on the cytoplasmic side of the inner membrane) and cytochrome b_2 (located in the intermembrane space), are both processed in two separate steps during transfer. The second processing step requires heme for cytochrome c_1, which has a covalently linked heme group, but not for cytochrome b_2, whose heme group is not covalently attached (Gasser *et al.*, 1982). Both precursors are processed to the apparent molecular weight of their intermediate forms by the protease found in the matrix. It was therefore suggested that the polypeptides initially extend into the matrix space during their transfer, so that a portion of the sequence can be removed by the matrix protease. Following this step the intermediate form would return to the intermembrane space side of the inner membrane and be processed again by a different enzyme (Gasser *et al.*, 1982). Both cytochrome c_1 and cytochrome b_2 require a membrane potential for the first part of the transfer process (Teintze *et al.*, 1982; Gasser *et al.*, 1982), and the processing of cytochrome c_1 from the intermediate form to the mature size also requires energy (Teintze *et al.*, 1982). It is possible that other proteins share this two-step processing mechanism, but their intermediates may be so short-lived that they have escaped detection.

Insertion into or transfer across the inner membrane evidently requires an intact membrane potential, even for proteins like the ADP/ATP carrier that are not proteolytically processed (Schleyer *et al.*, 1982). The energy dependence of the import of cytochrome b_2 into the intermembrane space further supports the idea that this enzyme makes a "detour" through the inner membrane to reach its functional location.

The least-understood step in the transfer process at this time is how the proteins get through or into the membrane after recognition by a receptor on the mitochondria. They cannot be "pushed" through by polypeptide chain elongation as they may be in cotranslational transport, so some feature of the protein sequence must result in this translocation. So far, however, no common structure has been found among the many imported mitochondrial proteins that would suggest how this might occur.

REFERENCES

Autor, A. P., 1982, Biosynthesis of mitochondrial manganese superoxide dismutase in *Saccharomyces cerevisiae*. Precursor form of mitochondrial superoxide dismutase made in the cytoplasm, *J. Biol. Chem.* **257**:2713–2718.

Blobel, G., and Dobberstein, B., 1975, Transfer of proteins across membranes, *J. Cell Biol.* **67**:835–851.

Böhni, P., Grasser, S., Leaver, C., and Schatz, G., 1980, A matrix-localized mitochondrial protease processing cytoplasmically-made precurors to mitochondrial proteins, in: *The Organization and Expression of the Mitochondrial Genome* (A. Kroon and C. Saccone, eds.) Elsevier, Amsterdam, pp. 423–433.

Chua, N.-H., and Schmidt, G. W., 1978, Post-translational transport into intact chloroplasts of a precursor to the small subunit of ribulose-1,5-bisphosphate carboxylase, *Proc. Natl. Acad. Sci. USA* **75**:6110–6114.

Chua, N.-H., and Schmidt, G. W., 1979, Transport of proteins into mitochondria and chloroplasts, *J. Cell Biol.* **81**:461–483.

Conboy, J. G., and Rosenberg, L. E., 1981, Post-translational uptake and processing of *in vitro* synthesized ornithine transcarbamoylase precursor by isolated rat liver mitochondria, *Proc. Natl. Acad. Sci. USA* **78**:3073–3077.

Conboy, J. G., Fenton, W. A., and Rosenberg, L. E., 1982, Processing of pre-ornithine transcarbamylase requires a zinc-dependent protease localized to the mitochondrial matrix, *Biochem. Biophys. Res. Commun.* **105**:1–7.

Côte, C., Solioz, M., and Schatz, G., 1979, Biogenesis of the cytochrome bc_1 complex of yeast mitochondria. A precursor form of the cytoplasmically made subunit V, *J. Biol. Chem.* **254**:1437–1439.

DuBois, R. N., Simpson, E. R., Tuckey, J., Lambeth, J. D., and Waterman, M. R., 1981, Evidence for a higher molecular weight precursor of cholesterol side-chain-cleavage cytochrome P-450 and induction of mitochondrial and cytosolic proteins by corticotropin in adult bovine adrenal cells, *Proc. Natl. Acad. Sci. USA* **78**:1028–1032.

Freitag, H., Neupert, W., and Benz, R., 1982a, Purification and characterization of a pore protein of the outer mitochondrial membrane from *Neurospora crassa*, *Eur. J. Biochem.* **123**:629–636.

Freitag, H., Janes, M., and Neupert, W., 1982b, Biosynthesis of mitochondrial porin and insertion into the outer mitochondrial membrane of *Neurospora crassa*, *Eur. J. Biochem.* **126**:197–202.

Gasser, S. M., Ohashi, A., Daum, G., Böhni, P. C., Gibson, J., Reid, G. A., Yonetani, T., and Schatz, G., 1982, Imported mitochondrial proteins cytochrome b_2 and cytochrome c_1 are processed in two steps, *Proc. Natl. Acad. Sci. USA* **79**:267–271.

Graven, S. N., Estrada-O., S., and Lardy, H. A., 1966, Alkali metal cation release and respiratory inhibition induced by nigericin in rat liver mitochondria, *Proc. Natl. Acad. Sci. USA* **56**:654–658.

Hallermayer, G., Zimmermann, R., and Neupert, W., 1977, Kinetic studies on the transport of cytoplasmically synthesized proteins into mitochondria in intact cells of *Neurospora crassa, Eur. J. Biochem.* **81**:523–532.

Harmey, M. A., and Neupert, W., 1979, Biosynthesis of mitochondrial citrate synthase in *Neurospora crassa, FEBS Lett.* **108**:385–389.

Heinrich, P. C., 1982, Proteolytic processing of polypeptides during the biosynthesis of subcellular structures, *Rev. Physiol. Biochem. Pharmacol.* **93**:115–187.

Heldt, H. W., Klingenberg, M., and Milovancev, M., 1972, Differences between the ATP/ADP ratios in the mitochondrial matrix and in the extramitochondrial space, *Eur. J. Biochem.* **30**:434–440.

Hennig, B., and Neupert, W., 1981, Assembly of cytochrome c. Apocytochrome c is bound to specific sites on mitochondria before its conversion to holocytochrome c, *Eur. J. Biochem.* **121**:203–212.

Hensgens, L. A. M., Grivell, L. A., Borst, P., and Bos, J. L., 1979, Nucleotide sequence of the mitochondrial gene for subunit 9 of yeast ATPase complex, *Proc. Natl. Acad. Sci. USA* **76**:1663–1667.

Jackson, R. C., and Blobel, G., 1977, Posttranslational cleavage of presecretory proteins with an extract of rough microsomes from dog pancreas containing signal peptidase activity, *Proc. Natl. Acad. Sci. USA* **74**:5598–5602.

Klingenberg, M., Aquila, H., and Riccio, P., 1979, Isolation of functional membrane proteins related to or identical with the ADP,ATP carrier of mitochondria, *Methods Enzymol.* **56**:407–414.

Korb, H., and Neupert, W., 1978, Biogenesis of cytochrome c in *Neurospora crassa.* Synthesis of apocytochrome c, transfer to mitochondria and conversion to holocytochrome c, *Eur. J. Biochem.* **91**:609–620.

Kreibich, G., Czako-Graham, M., Grebenau, R., Mok, W., Rodriguez-Boulan, E., and Sabatini, D. D., 1978, Characterization of the ribosomal binding site in rat liver rough microsomes: Ribophorins I and II, two integral membrane proteins related to ribosome binding, *J. Supramol. Struct.* **8**:279–302.

Lewin, A. S., Gregor, I., Mason, T. L., Nelson, N., and Schatz, G., 1980, Cytoplasmically made subunits of yeast mitochondrial F_1-ATPase and cytochrome c oxidase are synthesized as individual precursors, not as polyproteins, *Proc. Natl. Acad. Sci. USA* **77**:3998–4002.

Lloyd, D., 1974, *The Mitochondria of Microorganisms,* Academic Press, London, pp. 82–158.

McAda, P. C., and Douglas, M. G., 1982, A neutral metallo endoprotease involved in the processing of an F_1-ATPase subunit precursor in mitochondria, *J. Biol. Chem.* **257**:3177–3182.

Maccecchini, M.-L., Rudin, Y., Blobel, G., and Schatz, G., 1979a, Import of proteins into mitochondria: Precursor forms of the extramitochondrially made F_1-ATPase subunits in yeast, *Proc. Natl. Acad. Sci. USA* **76**:343–347.

Maccecchini, M.-L., Rudin, Y., and Schatz, G., 1979b, Transport of proteins across the mitochondrial outer membrane. A precursor form of the cytoplasmically made intermembrane enzyme cytochrome c peroxidase, *J. Biol. Chem.* **254**:7468–7471.

Macino, G., and Tzagoloff, A., 1979, Assembly of the mitochondrial membrane system. The DNA sequence of a mitochondrial ATPase gene in *Saccharomyces cerevisiae, J. Biol. Chem.* **254**:4617–4623.

Matsuura, S., Arpin, M., Hannum, C., Margoliash, E., Sabatini, D. D., and Morimoto, T., 1981, *In vitro* synthesis and post-translational uptake of cytochrome c into isolated mitochondria: Role of a specific addressing signal in the apocytochrome, *Proc. Natl. Acad. Sci. USA* **78**:4368–4372.

Meyer, D. I., and Dobberstein, B., 1980, Identification and characterization of a membrane component essential for the translocation of nascent protein across the membrane of the endoplasmic reticulum, *J. Cell Biol.* **87**:503–508.

Meyer, D. I., Krause, E., and Dobberstein, B., 1982, Secretory protein translocation across membranes: the role of the "docking protein," *Nature* **297**:647–650.

Michel, R., Wachter, E., and Sebald, W., 1979, Synthesis of a larger precursor for the proteolipid subunit of the mitochondrial ATPase complex of *Neurospora crassa* in a cell-free wheat germ system, *FEBS Lett.* **101**:373–376.

Mihara, K., and Blobel, G., 1980, The four cytoplasmically made subunits of yeast mitochondrial cytochrome *c* oxidase are synthesized individually and not as a polyprotein, *Proc. Natl. Acad. Sci. USA* **77**:4160–4164.

Mihara, K., Omura, T., Harano, T., Brenner, S., Fleischer, S., Rajagopalan, K. V., and Blobel, G., 1982, Rat liver L-glutamate dehydrogenase, D-β-hydroxybutyrate dehydrogenase, and sulfite oxidase are each synthesized as larger precursors by cytoplasmic free polysomes, *J. Biol. Chem.* **257**:3355–3358.

Miura, S., Mori, M., Amaya, Y., and Tatibana, M., 1982, A mitochondrial protease that cleaves the precursor of ornithine carbamoyltransferase. Purification and properties, *Eur. J. Biochem.* **122**:641–647.

Mori, M., Miura, S., Tatibana, M., and Cohen, P. P., 1979, Cell-free synthesis and processing of a putative precursor for mitochondrial carbamyl phosphate synthetase I of rat liver, *Proc. Natl. Acad. Sci. USA* **76**:5071–5075.

Mori, M., Miura, S., Tatibana, M., and Cohen, P. P., 1980, Processing of a putative precursor of rat liver ornithine transcarbamylase, a mitochondrial matrix enzyme, *J. Biochem.* **88**:1829–1836.

Morita, T., Miura, S., Mori, M., and Tatibana, M., 1982, Transport of the precursor for rat-liver ornithine carbamoyltransferase into mitochondria *in vitro*, *Eur. J. Biochem.* **122**:501–509.

Nabi, N., and Omura, T., 1980, *In vitro* synthesis of adrenodoxin and adrenodoxin reductase: Existence of a putative large precursor form of adrenodoxin, *Biochem. Biophys. Res. Commun.* **97**:680–686.

Nabi, N., Kominami, S., Takemori, S., and Omura, T., 1980, *In vitro* synthesis of mitochondrial cytochromes P-450 (scc) and P-450 (11-β) and microsomal cytochrome P-450 (c-21) by both free and bound polysomes isolated from bovine adrenal cortex, *Biochem. Biophys. Res. Commun.* **97**:687–693.

Nelson, N., and Schatz, G., 1979, Energy-dependent processing of cytoplasmically made precursors to mitochondrial proteins, *Proc. Natl. Acad. Sci. USA* **76**:4365–4369.

Neupert, W., and Schatz, G., 1981, How proteins are transported into mitochondria, *Trends Biochem. Sci.* **6**:1–4.

Okada, Y., Frey, A. B., Guenther, T. M., Oesch, F., Sabatini, D. D., and Kreibich, G., 1982, Studies on the biosynthesis of microsomal membrane proteins. Site of synthesis and mode of insertion of cytochrome b_5, cytochrome b_5 reductase, cytochrome P-450 reductase and expoxide hydrolase, *Eur. J. Biochem.* **122**:393–402.

Palade, G., 1975, Intracellular aspects of the process of protein synthesis, *Science* **189**:347–358.

Pressman, B. C., 1976, Biological applications of ionophores, *Annu. Rev. Biochem.* **45**:501–530.

Raymond, Y., and Shore, G. C., 1979, The precursor for carbamyl phosphate synthetase is transported to mitochondria via a cytosolic route, *J. Biol. Chem.* **254**:9335–9338.

Raymond, Y., and Shore, G. C., 1980, Kinetics of uptake and processing of the precursor for carbamyl phosphate synthetase by mitochondria *in vivo* and *in vitro*, *J. Cell Biol.* **87**:MC1424 (abstr.).

Sabatini, D. D., Kreibich, G., Morimoto, T., and Adesnik, M., 1982, Mechanisms for the incorporation of proteins in membranes and organelles, *J. Cell Biol.* **92**:1–22.

Sakakibara, R., Huynh, Q. K., Nishida, Y., Watanabe, T., and Wade, H., 1980, *In vitro* synthesis of glutamic oxaloacetic transaminase isozymes of rat liver, *Biochem. Biophys. Res. Commun.* **95**:1781–1788.

Scarpa, A., 1979, Transport across mitochondrial membranes, in: *Membrane Transport in Biology. II. Transport Across Single Biological Membranes* (D. C. Tosteson, ed.), Springer-Verlag, Berlin, pp. 263–355.

Schatz, G., 1979, How mitochondria import proteins from the cytoplasm, *FEBS Lett.* **103**:201–211.

Schatz, G., and Mason, T. L., 1974, The biosynthesis of mitochondrial proteins, *Annu. Rev. Biochem.* **43**:51–87.

Schleyer, M., Schmidt, B., and Neupert, W., 1982, Requirement of a membrane potential for the posttranslational transfer of proteins into mitochondria, *Eur. J. Biochem.* **125**:109–116.

Schmelzer, E., and Heinrich, P. C., 1980, Synthesis of a larger precursor for the subunit IV of rat liver cytochrome *c* oxidase in a cell-free wheat germ system, *J. Biol. Chem.* **255**:7503–7506.

Sebald, W., and Wachter, E., 1978, Amino acid sequence of the putative protonophore of the energy-transducing ATPase complex, in: *29th Moosbach Colloquium: Energy Conservation in Biological Membranes* (G. Schäfer and M. Klingenberg, eds.), Springer-Verlag, Berlin, pp. 228–263.

Shore, G. C., Carignan, P., and Raymond, Y., 1979, *In vitro* synthesis of a putative precursor to the mitochondrial enzyme, carbamyl phosphate synthetase, *J. Biol. Chem.* **254**:3141–3144.

Sonderegger, P., Jaussi, R., and Christen, P., 1980, Cell-free synthesis of a putative precursor of mitochondrial aspartate aminotransferase with higher molecular weight, *Biochem. Biophys. Res. Commun.* **94**:1256–1260.

Sonderegger, P., Jaussi, R., Christen, P., and Gehring, H., 1982, Biosynthesis of aspartate aminotransferase. Both the higher molecular weight precursor of mitochondrial aspartate aminotransferase and the cytosolic isoenzyme are synthesized on free polysomes, *J. Biol. Chem.* **257**:3339–3345.

Stigall, D. L., Galante, Y. M., and Hatefi, Y., 1979, Preparation and properties of complex V, *Methods Enzymol.* **55**:308–315.

Subik, J., Kolarov, J., and Kovac, L., 1974, Bongkrekic acid sensitivity of respiration-deficient mutants and of petite negative species of yeast, *Biochim. Biophys. Acta* **357**:453–456.

Teintze, M., Slaughter, M., Weiss, H., and Neupert, W., 1982, Biogenesis of mitochondrial ubiquinol:cytochrome *c* reductase (cytochrome bc_1 complex). Precursor proteins and their transfer into mitochondria, *J. Biol. Chem.* **257**:10364–10371.

Tzagoloff, A., Macino, G., and Sebald, W., 1979, Mitochondrial genes and translation products, *Annu. Rev. Biochem.* **48**:419–441.

Walter, P., and Blobel, G., 1981a, Translocation of proteins across the endoplasmic reticulum. II. Signal recognition protein (SRP) mediates the selective binding to microsomal membranes of *in-vitro*-assembled polysomes synthesizing secretory protein, *J. Cell Biol.* **91**:551–556.

Walter, P., and Blobel, G., 1981b, Translocation of proteins across the endoplasmic reticulum. III. Signal recognition protein (SRP) causes signal sequence-dependent and site-specific arrest of chain elongation that is released by microsomal membranes, *J. Cell Biol.* **91**:557–561.

Walter, P., Jackson, R. C., Marcus, M. M., Lingappa, V. R., and Blobel, G., 1979, Tryptic dissection and reconstitution of translocation activity for nascent presecretory proteins across microsomal membranes, *Proc. Natl. Acad. Sci. USA* **76**:1795–1799.

Walter, P., Ibrahimi, I., and Blobel, G., 1981, Translocation of proteins across the endoplasmic reticulum. I. Signal recognition protein (SRP) binds to *in-vitro*-assembled polysomes synthesizing secretory protein, *J. Cell Biol.* **91**:545–550.

Weiss, H., and Kolb, H. J., 1979, Isolation of mitochondrial succinate:ubiquinone reductase, cytochrome *c* reductase and cytochrome *c* oxidase from *Neurospora crassa* using nonionic detergent, *Eur. J. Biochem.* **99**:139–149.

Wickner, W., 1980, Assembly of proteins into membranes, *Science* **210**:861–862.

Yamauchi, K., Hayashi, N., and Kikuchi, G., 1980a, Translocation of δ-aminolevulinate synthase from the cytosol to the mitochondria and its regulation by hemin in the rat liver, *J. Biol. Chem.* **255**:1746–1751.

Yamauchi, K., Hayashi, N., and Kikuchi, G., 1980b, Cell-free synthesis of rat liver δ-aminolevulinate synthase and possible occurrence of processing of the enzyme protein in the course of its translocation from the cytosol into the mitochondrial matrix, *FEBS Lett.* **115**:15–18.

Zimmermann, R., and Neupert, W., 1980, Transport of proteins into mitochondria. Posttranslational transfer of ADP/ATP carrier into mitochondria *in vitro, Eur. J. Biochem.* **109**:217–229.

Zimmermann, R., Paluch, U., and Neupert, W., 1979a, Cell-free synthesis of cytochrome *c*, *FEBS Lett.* **108**:141–146.

Zimmermann, R., Paluch, U., Sprinzl, M., and Neupert, W., 1979b, Cell-free synthesis of the mitochondrial ADP/ATP carrier protein of *Neurospora crassa, Eur. J. Biochem.* **99**:247–252.

Zimmermann, R., Hennig, B., and Neupert, W., 1981, Different transport pathways of individual precursor proteins in mitochondria, *Eur. J. Biochem.* **116**:455–460.

Zwizinski, C., and Wickner, W., 1980, Purification and characterization of leader (signal) peptidase from *Escherichia coli, J. Biol. Chem.* **255**:7973–7977.

4

POLYPEPTIDE-HORMONE-INDUCED RECEPTOR CLUSTERING AND INTERNALIZATION

J. Schlessinger, A. B. Schreiber, T. A. Libermann, I. Lax, A. Avivi, and Y. Yarden

1. INTRODUCTION

Many cellular processes are mediated and controlled by polypeptide hormones and growth factors (Carpenter and Cohen, 1979; Levi-Montalcini and Angeletti, 1968; Gospodarowicz and Moran, 1976). The first step in the action of all polypeptide hormones and growth factors is specific binding to their membrane receptors. This step initiates a broad range of biochemical events: early effects on cell metabolism and delayed effects on the proliferation and differentiation of target cells.

It is possible to list several common features in the action and regulation of nearly all polypeptide growth factors:

1. All of them bind to their specific membrane receptors (Gospodarowicz and Moran, 1976; Bradshaw, 1978).
2. All of them induce receptor loss through a process called down-regulation (Kahn, 1976; Bradshaw, 1978; Aharonov *et al.*, 1978; Carpenter and Cohen, 1979).
3. The polypeptide hormones and growth factors are degraded subsequent to their binding to target cells (Kahn, 1976; Bradshaw, 1978; Carpenter and Cohen, 1979).

J. Schlessinger, A. B. Schreiber, T. A. Libermann, I. Lax, A. Avivi, and Y. Yarden ● Department of Chemical Immunology, The Weizmann Institute of Science, Rehovot 76100, Israel.

4. Several growth factors—such as insulin (Goldfine et al., 1977), nerve growth factor (NGF) (Andres et al., 1977; Yankner and Shooter, 1979), and epidermal growth factor (EGF) (Johnson et al., 1980)— bind to nuclear receptors.

Thus, the mode of action of growth factors involves many complex biochemical processes. Moreover, it seems that the growth factors probably do not utilize a single "second messenger" (e.g., cAMP) for their action. Alternatively, growth factors could activate an array of temporally synchronized biochemical reactions that lead to the activation of the various early and delayed responses as well as the induction of receptor down-regulation and hormone degradation. Consequently, it is obvious that, in order to understand the mechanism of action of growth factors, it is necessary to adapt several strategies involving many methodologies. Thus, it is important to characterize for every hormone–receptor system the following points:

1. Characterization of the membrane receptor for the growth factor, its subunit composition, chemical nature (i.e., enzyme, ion channel), and possible interactions with other membrane molecules.
2. Development of a reliable cell-free system that responds specifically to the growth factor and allows the in vitro analysis of the biochemical processes that are induced by the growth factor.
3. Detailed morphological analysis by electron microscopy of the localization of radiolabeled or electron-dense-labeled hormone–receptor complexes in cells under various physiological conditions. It is also important to analyze the mobility and localization of the occupied receptors on living cells in order to characterize the dynamic properties of the receptors in situ.

Only the combination of these three strategies will provide a detailed picture that will eventually lead to the understanding of the mode of action of growth factors.

This review is divided into two principal sections. The next section describes the process of polypeptide-hormone-induced receptor clustering and internalization. This seems to be a general pathway for receptor-mediated endocytosis of polypeptide hormones, growth factors, and other serum proteins. In the following section we review the EGF–receptor system. We describe in detail the dynamic properties of the EGF–receptor complex in situ and also provide a detailed summary of other studies that employ each of the three strategies described above.